145
Structure and Bonding

Series Editor: D. M. P. Mingos

For further volumes:
http://www.springer.com/series/430

Structure and Bonding

Series Editor: D. M. P. Mingos

Recently Published and Forthcoming Volumes

Structure-Property Relationships in Non-Linear Optical Crystals II
Volume Editors: Xin-Tao Wu, Ling Chen
Vol. 145, 2012

Structure-Property Relationships in Non-Linear Optical Crystals I
Volume Editors: Xin-Tao Wu, Ling Chen
Vol. 144, 2012

Molecular Electronic Structures of Transition Metal Complexes II
Volume Editors: D. M. P. Mingos, Peter Day, Jens Peder Dahl
Vol. 143, 2012

Molecular Electronic Structures of Transition Metal Complexes I
Volume Editors: D. M. P. Mingos, Peter Day, Jens Peder Dahl
Vol. 142, 2012

Fuel Cells and Hydrogen Storage
Volume Editors: Andrew Bocarsly, D. M. P. Mingos
Vol. 141, 2011

Zintl Ions
Principles and Recent Developments
Volume Editor: Thomas F. Fässler
Vol. 140, 2011

Zintl Phases
Principles and Recent Developments
Volume Editor: Thomas F. Fässler
Vol. 139, 2011

Inorganic 3D Structures
Volume Editor: Angel Vegas
Vol. 138, 2011

Molecular Catalysis of Rare-Earth Elements
Volume Editor: Peter W. Roesky
Vol. 137, 2010

Metal-Metal Bonding
Volume Editor: Gerard Parkin
Vol. 136, 2010

Functional Phthalocyanine Molecular Materials
Volume Editor: Jianzhuang Jiang
Vol. 135, 2010

Data Mining in Crystallography
Volume Editors: Hofmann, D. W. M., Kuleshova, L. N.
Vol. 134, 2010

Controlled Assembly and Modification of Inorganic Systems
Volume Editor: Wu, X.- T.
Vol. 133, 2009

Molecular Networks
Volume Editor: Hosseini, M. W.
Vol. 132, 2009

Molecular Thermodynamics of Complex Systems
Volume Editors: Lu, X., Hu, Y.
Vol. 131, 2009

Contemporary Metal Boron Chemistry I
Volume Editors: Marder, T. B., Lin, Z.
Vol. 130, 2008

Recognition of Anions
Volume Editor: Vilar, R.
Vol. 129, 2008

Liquid Crystalline Functional Assemblies and Their Supramolecular Structures
Volume Editor: Kato, T.
Vol. 128, 2008

Organometallic and Coordination Chemistry of the Actinides
Volume Editor: Albrecht-Schmitt, T. E.
Vol. 127, 2008

Halogen Bonding
Fundamentals and Applications
Volume Editors: Metrangolo, P., Resnati, G.
Vol. 126, 2008

Structure-Property Relationships in Non-Linear Optical Crystals II

The IR Region

Volume Editors:
Xin-Tao Wu · Ling Chen

With contributions by

G.-C. Guo · S.-P. Guo · X.-M. Jiang · C.-Y. Tu · K.-C. Wu · L.-M. Wu · X.-T. Wu · H.-Y. Zeng · M.-J. Zhang · H.-J. Zhao

Springer

Editors
Xin-Tao Wu
Ling Chen

Fujian Institute of Research on the
Structure of Matter (FJIRSM)
Chinese Academy of Sciences
Fuzhou, Fujian
China

ISSN 0081-5993 ISSN 1616-8550 (electronic)
ISBN 978-3-642-29620-8 ISBN 978-3-642-29621-5 (eBook)
DOI 10.1007/978-3-642-29621-5
Springer Heidelberg New York Dordrecht London

Library of Congress Control Number: 2012938650

Printed on acid-free paper

Springer is part of Springer Science+Business Media (www.springer.com)

Series Editor

Prof. D. Michael P. Mingos

Inorganic Chemistry Laboratory
Oxford University
South Parks Road
Oxford OX1 3QR, UK
michael.mingos@st-edmund-hall.oxford.ac.uk

Volume Editors

Xin-Tao Wu
Ling Chen

Fujian Institute of Research on the
Structure of Matter (FJIRSM)
Chinese Academy of Sciences
Fuzhou, Fujian
China

Editorial Board

Structure and Bonding
Also Available Electronically

Structure and Bonding is included in Springer's eBook package *Chemistry and Materials Science*. If a library does not opt for the whole package the book series may be bought on a subscription basis. Also, all back volumes are available electronically.

For all customers who have a standing order to the print version of *Structure and Bonding*, we offer the electronic version via SpringerLink free of charge.

If you do not have access, you can still view the table of contents of each volume and the abstract of each article by going to the SpringerLink homepage, clicking on "Chemistry and Materials Science," under Subject Collection, then "Book Series," under Content Type and finally by selecting *Structure and Bonding*.

You will find information about the

- Editorial Board
- Aims and Scope
- Instructions for Authors
- Sample Contribution

at springer.com using the search function by typing in *Structure and Bonding*.

Color figures are published in full color in the electronic version on SpringerLink.

Aims and Scope

The series *Structure and Bonding* publishes critical reviews on topics of research concerned with chemical structure and bonding. The scope of the series spans the entire Periodic Table and addresses structure and bonding issues associated with all of the elements. It also focuses attention on new and developing areas of modern structural and theoretical chemistry such as nanostructures, molecular electronics, designed molecular solids, surfaces, metal clusters and supramolecular structures. Physical and spectroscopic techniques used to determine, examine and model structures fall within the purview of *Structure and Bonding* to the extent that the focus is on the scientific results obtained and not on specialist information concerning the techniques themselves. Issues associated with the development of bonding models and generalizations that illuminate the reactivity pathways and rates of chemical processes are also relevant.

The individual volumes in the series are thematic. The goal of each volume is to give the reader, whether at a university or in industry, a comprehensive overview of an area where new insights are emerging that are of interest to a larger scientific audience. Thus each review within the volume critically surveys one aspect of that topic and places it within the context of the volume as a whole. The most significant developments of the last 5 to 10 years should be presented using selected examples to illustrate the principles discussed. A description of the physical basis of the experimental techniques that have been used to provide the primary data may also be appropriate, if it has not been covered in detail elsewhere. The coverage need not be exhaustive in data, but should rather be conceptual, concentrating on the new principles being developed that will allow the reader, who is not a specialist in the area covered, to understand the data presented. Discussion of possible future research directions in the area is welcomed.

Review articles for the individual volumes are invited by the volume editors.

In references *Structure and Bonding* is abbreviated *Struct Bond* and is cited as a journal.

Impact Factor in 2010: 4.659; Section "Chemistry, Inorganic & Nuclear":
Rank 4 of 43; Section "Chemistry, Physical": Rank 25 of 127

In memory of the 100th birthday of Prof. Jia-Xi Lu (1915–2015)

Dedicated to Fujian Institute of Research on the Structure of Matter, Chinese Academy of Sciences, on the occasion of its 55th anniversary (1960–2015)

Preface

Nonlinear optics (NLO) is a branch of optics which describes the behavior of light in *nonlinear media*, that is, media in which the dielectric polarization P responds nonlinearly to the electric field E of the light. This nonlinearity is generally only observed at very high light intensities such as those provided by pulsed lasers. The discovery of lasers in the 1960s revolutionized the study of nonlinear optical materials. Nonlinear optics gives rise to a host of optical phenomena based on the frequency mixing processes such as second, third, and higher harmonic generation, sum and difference frequency generation, optical parametric amplification, oscillation and generation, and optical rectification. Nonlinear optics is now a multidisciplinary field of study, which incorporates not only fundamental studies on the theoretical basis of the phenomenon, but also the fabrication of nonlinear optical devices and the discovery of new nonlinear optical crystalline materials. New crystalline materials are of great importance to the development of the field of nonlinear optics, especially since the continuous discovery of novel, effective, high quality of nonlinear optical crystals provides great opportunities for the development of nonlinear devices, which promote the wide application of laser frequency conversion techniques in many fields, such as laser communication, laser medical treatment, laser nuclear fusion, laser distance measurement, and fundamental spectroscopic research.

In 1960, the founder of Fujian Institute of Research on the Structure of Matter, Chinese Academy of Sciences (FJIRSM), Prof. Jia-Xi Lu, initiated and motivated a comprehensive research base in China specializing in structural chemistry and new crystalline materials. After more than 50 years of untiring efforts, the Institute remains devoted to this area of fundamental research and the development manufacturing and marketing of a wide range of nonlinear optical crystals. CASTECH, the company founded by FJIRSM in 1988, has become the leading supplier for LBO, BBO, and numerous other nonlinear optical crystals in the world.

We have organized two volumes to summarize some of the recent progress focusing on structure–property relationships in nonlinear optical crystals. Volume 145 is devoted to the IR region with four review-type chapters that were written by five of our faculty members who are the leading scientists in their field.

Prof. G.-C. Guo reviews the crystal growth of large crystals and new mid-infrared second-order NLO pnictide, chalcogenide, and halide materials. Prof. K.-C. Wu discusses the simulation and design of IR second-order NLO metal cluster compounds, Prof. C.-Y. Tu describes the recent development of stimulated Raman scattering (SRS) and SRS self-frequency conversion laser crystals, and Profs. L.-M. Wu and X.-T. Wu review the exploration of new second-order NLO compounds containing main group elements.

We wish to acknowledge our deep appreciations for all of the authors who unselfishly spent their most precious time writing contributions for this volume. We also gratefully acknowledge the continuous support from CAS, NNSF, MOST, and Fujian Province.

Fuzhou
April 2012

Xin-Tao Wu and Ling Chen

Contents

Struct Bond (2012) 145: 1–44
DOI: 10.1007/430_2011_72

Published online: 3 March 2012

Large Crystal Growth and New Crystal Exploration of Mid-Infrared Second-Order Nonlinear Optical Materials

Xiao-Ming Jiang, Sheng-Ping Guo, Hui-Yi Zeng, Ming-Jian Zhang, and Guo-Cong Guo

Abstract Mid-IR second-order NLO crystal is indispensable in the frequency conversion applications in the mid-IR region. Compared with DUV and UV/Vis/near-IR NLO crystals, practical mid-IR NLO crystals are relatively rare, and many of them are still at the stage of laboratory research. This chapter reviews the recent progress on the mid-IR NLO crystals, which mainly includes growing the classical mid-IR NLO crystals into large high-quality ones or into quasi-phase-matching structures that are suitable for the laser devices by various growth methods and exploring new potential mid-IR NLO crystals by introducing new design and synthesis strategies. Recent mid-IR NLO crystals can be divided into four categories, i.e., classical binary and ternary metal pnictides and chalcogenides, quaternary metal chalcogenides, binary and ternary metal halides, and different-bond-type hybrid compounds that contain at least two types of obviously different chemical bonds in the crystal structures. Metal pnictides and chalcogenides have got much attention on growing large crystals. Different-bond-type hybrid is a new family of mid-IR NLO materials, and many of them were found in the last decade. In metal halide system, both progress in growing large crystals and discovering new ones have been made.

Keywords Chalcogenide · Halide · Large crystal · Mid-infrared · Nonlinear optical · Pnictide

---

X.-M. Jiang, S.-P. Guo, H.-Y. Zeng, M.-J. Zhang,
and Guo-Cong Guo (✉)
State Key Laboratory of Structural ChemistryFujian Institute of Research on the Structure of Matter, Chinese Academy of Sciences, Fuzhou, Fujian 350002, People's Republic of China
e-mail: gcguo@fjirsm.ac.cn

Contents

Abbreviations

BS	Bridgman–Stockbarger
BTPs	Bicapped trigonal prisms
DUV	Deep ultraviolet
HRGC	$RbGeCl_3{\cdot}x(H_2O)$
KDP	KH_2PO_4
KTP	$KTiOPO_4$
LIDT	Laser-induced damage threshold
Mid-IR	Mid-infrared
NCS	Noncentrosymmetric
NLO	Nonlinear optical
QPM	Quasi-phase-matching
SHG	Second-harmonic generation
Vis	Visible
ZGP	$ZnGeP_2$

1 Introduction

It is well known that laser plays a very important role in the modern society; too many applications of laser in industry, medicine, military, and scientific research such as spectroscopy, microscopy, photochemistry, material processing, nuclear fusion, and so on, have come true [1, 2]. In order to extend laser frequency ranges, the important method, frequency conversion is often used, in which an efficient and stable frequency-shifting device, nonlinear optical (NLO) crystal is indispensable [3].Generally, we always refer to NLO crystals to the second-order NLO crystals if no other instructions are provided, and it is like that in this chapter. Only crystallographically noncentrosymmetric (NCS) compounds can have second-order NLO performance according to the Neumann principle. A good NLO crystal should possess some necessary attributes for second-harmonic generation (SHG) applications, for example, large nonlinear coefficient, wideband spectral transparency, low absorption loss at convenient pump-laser wavelengths, high laser-damage threshold, reasonable chemical stability, good mechanical robustness, high thermal conductivity, low thermal index change, and birefringently phase-matched over an appropriate wavelength regime [4, 5]. Up to now, lots of practical NLO crystals have been found and applied in deep ultraviolet (DUV, <300 nm) and ultraviolet/visible/near-infrared (UV/Vis/near-IR, 0.3–3 μm) region (Table 1), but they cannot be used in mid-infrared region especially two atmospheric transparent windows, 3–5 and 8–14 μm, due to strong absorption result from their oxygen content [6–8, 16]. In this chapter, mid-IR NLO crystals only refer to those inorganic crystals without oxygen content.

Up to now, practical mid-IR NLO crystals are relatively rare [27–29], and many of them are still at the stage of laboratory research when compared with DUV and UV/Vis/near-IR NLO crystals. To find new NLO materials with wide transparency range in IR region, high damage resistance to the incident laser beam and good chemical stability to the moisture and oxygen are still the charming and challenging subjects in this field.

During the last decade, mainly two types of work on the mid-IR NLO crystals have been done by our and other groups. One type is growing the so-called classical mid-IR NLO crystals, which will be described in detail afterward, into large high-quality ones that are suitable for the laser devices, mainly by the Bridgman–Stockbarger (BS) growth method [30, 31] and MBE growth method [32] that is used to introduce quasi-phase-matching (QPM) structures for optical isotropic crystals [33–35]. The classical mid-IR NLO crystals [36, 37] in this chapter refer to those having a relative long history than others, which mainly include binary (II, III)–VI- and III–V-type

Table 1 Practical DUV (<300 nm) and UV/Vis/near-IR (0.3–3 μm) NLO crystals

DUV NLO crystals	UV/Vis/near-IR NLO crystals
KBBF family: $KBe_2BO_3F_2$(KBBF) [6], $RbBe_2BO_3F_2$(RBBF) [7], $BaAlBO_3F_2$(BABF) [8]	*Borates*: β-BaB_2O_4(BBO) [9], LiB_3O_5(LBO) [10], CsB_3O_5(CBO) [11], $CsLiB_6O_{10}$(CLBO) [12], $La_2CaB_{10}O_{19}$(LCB) [13], $Na_3La_9O_3(BO_3)_8$(NLBO) [14], $Bi_2ZnOB_2O_6$(BZB) [15]
SBBO family [16]: $Sr_2Be_2B_2O_7$(SBBO), $Ba_2Be_2B_2O_7$(TBO), $BaAl_2B_2O_7$(BABO), $K_2Al_2B_2O_7$(KABO)	*Phosphates* [17]: KH_2PO_4(KDP), $NH_4H_2PO_4$(ADP), KD_2PO_4(DKDP), $KTiOPO_4$(KTP) [18]
Borophosphates [19]: β-Zn_3BPO_7(ZBP), $Mg_xZn_{3-x}BPO_7$(MZBP), $BaBPO_5$(BBP), $SrBPO_5$(SBP), BPO_4 [20]	*Niobates*: $LiNbO_3$(LN) [21], MgO: $LiNbO_3$(MgLN) [22], $KNbO_3$(KN) [23]
Borates: $YAl_3(BO_3)_4$ (YAB) [24], etc.	*Other oxysalts*: $KTiOAsO_4$(KTA) [25], α-$LiIO_3$ [26], etc.

semiconductor compounds (GaSe, GaAs, etc.), ternary II–IV–V_2- and I–III–VI_2-type chalcopyrite-structure compounds ($AgGaS_2$, $ZnGeP_2$, etc.), and some other ternary metal chalcogenides ($BaGa_4S_7$, $BaGa_4Se_7$, etc.). Quaternary diamond-like metal chalcogenides ($AgGaGeS_4$, $AgGaGe_5Se_{12}$, etc.) included in quaternary metal chalcogenides system and binary and ternary metal halides ($HgBr_2$, $BaMgF_4$, etc.) have entered the stage of growing large crystals and should also belong to the classical mid-IR NLO crystals, although a few new mid-IR NLO crystals are still found in the systems recently. Hydrothermal, aqueous, and Czochralski growth methods [38] are used to grow large mid-IR NLO metal halides. Although classical mid-IR NLO crystals posses many advantages, many of them have some shortcomings more or less, which hamper their wide applications. For example, GaSe has wide transparency range, high SHG efficiency, large birefringence, and high thermal conductivity but is highly cleavable and notoriously difficult to cut and polish [36].

The other type of work is discovering new crystals and new systems with good mid-IR NLO performances, like alkali-metal-based quaternary chalcogenides [$A_2Hg_3M_2S_8$ (A = K, Rb; M = Ge, Sn)] [39] and other quaternary metal chalcogenides ($ZnY_6Si_2S_{14}$, $Al_xDy_3(Si_yAl_{1-y})S_7$, $Al_{0.33}Sm_3SiS_7$, etc.) [40]. Some novel design and synthesis strategies are adopted to discover new mid-IR NLO crystals, for example, Kanatzidis's group adopted alkali metal polychalcogenide flux method [41] to synthesize metal thiotellurites [AAg_2TeS_6 (A = Rb, Cs)] [42] and chalcopnictates ($AZrPSe_6$ (A = K, Rb, Cs), etc.) [43]. Halasyamani's group discovered that thioborate system combining the favorable transparency and non-linearity of sulfur-based NLO materials and the high damage thresholds of borates are a new class of mid-IR NLO materials, such as $Zn_xBa_2B_2S_{5+x}$ ($x \approx 0.2$) [44]. Another special strategy of introducing polar polyanions and polycations in one compound simultaneously is first proposed by our group to design and synthesize mid-IR NLO crystals via supramolecular chemistry and crystal engineering methods, like metal pnictidohalides and chalcohalides [$(Hg_6P_3)(In_2Cl_9)$, $(Hg_8As_4)(Bi_3Cl_{13})$, etc.] [45]. These metal thiotellurites, chalcopnictates, thioborates, pnictidohalides, and chalcohalides have a common feature that they possess two obviously different types of bonds simultaneously in one compound; we call them different-bond-type hybrid compounds and will give a detailed description in the relevant chapter.

The reported main mid-IR NLO material systems and relevant crystals are listed in Table 2. Their achievements, which have been made in the past decade, will be reviewed in detail in the rest chapters. For the classical mid-IR NLO crystals, we will focus on the improving methods used to overcome some drawbacks which hamper their wide applications and growth techniques adopted to grow them into large high-quality crystals or QPM structures. For the discovered new potential mid-IR NLO crystals, mainly their crystal structures and mid-IR NLO properties, such as optical transparent region, SHG efficiency, laser-damage threshold, and so on, will get our attention.

Table 2 Practical and potential mid-IR NLO crystals

Classical binary and ternary metal pnictides and chalcogenides	Binary (II, III)–VI and III–V semiconductors	GaSe, GaAs, CdSe, HgS, and their doping materials [36, 37]
	II–IV–V2 and I–III–VI2 chalcopyrites	$AgGaX_2$ (X = S, Se, Te), $AgInX_2$ (X = S, Se), $LiMX_2$ (M = Al, Ga, In; X = S, Se, Te), $CuMX_2$ (M = Al, Ga; X = S, Se), $CuInS_2$, $ZnMP_2$ (M = Si, Ge), $CdSiX_2$ (X = P, As [46]), $CdGeAs_2$, and their mixing materials [36, 37]
	Other ternary metal chalcogenides	$BaGa_4X_7$ (X = S, Se) [47, 48], Ag_3QS_3 (Q = As, Sb) [36, 37], $HgGa_2S_4$ [49], $Cd_xHg_{1-x}Ga_2S_4$ [50], $ZnIn_2Te_4$ [51], $Pb_2GeS(e)_4$ [52]
Quaternary metal chalcogenides	Quaternary diamond-like metal chalcogenides	$AgGaGe_nS(e)_{2(n+1)}$ [53], $Li_2Ga_2GeS_6$ [54], Li_2CdGeS_4, Li_2CdSnS_4 [55]
	Quaternary alkali-metal-based chalcogenides	α- and β-$A_2Hg_3M_2S_8$ (A = K, Rb; M = Ge, Sn) [39], $Na_{0.5}Pb_{1.75}GeS_4$ [56]
	Other quaternary metal semiconductors	$ZnY_6Si_2S_{14}$, $Al_{0.50}Dy_3(Si_{0.50}Al_{0.50})S_7$, $Al_{0.38}Dy_3(Si_{0.85}Al_{0.15})S_7$, and $Al_{0.33}Sm_3SiS_7$ [40], $La_3CuGeSe_7$ [57], β-$LaGaS_3$ [58]
Binary and ternary metal halides		$HgBr_2$ [59], SbF_3 [60], Tl_nMX_{n+2} (n = 3, 4; M = Pb, Hg; X = halogen) [61], $BaMF_4$ (M = Mg, Mn, Co, Ni, and Zn) [62],$SrAlF_5$ [38], ABX_3 (A = alkali metal; B = Ge, Cd, Hg; X = halogen) [63], $CsGe(Br_xCl_{1-x})_3$ [64], $Cs_2Hg_3I_8$ [65], $NaSb_3F_{10}$ [66], Na_2SbF_5 [67], $NaLaF_4$ [68]
Different-bond-type hybrid compounds	Metal thiotellurites, chalcopnictates, and thioborates	$RbAg_2TeS_6$ and $CsAg_2TeS_6$ [42]; Tl_3AsSe_3 [69]; $LiAsS_2$ and $Li_{0.6}Na_{0.4}AsS_2$ [70]; β-$LiAsSe_2$, β-$Li_{0.2}Na_{0.8}AsSe_2$, γ-$Li_{0.2}Na_{0.8}AsSe_2$, and γ-$NaAsSe_2$ [71]; $K_3Ta_2AsS_{11}$ and $Rb_3Ta_2AsS_{11}$ [72]; $K_2P_2Se_6$ [73]; $APSe_6$ (A = K, Rb) [74]; $Cs_5P_5Se_{12}$ [75]; $AZrPSe_6$ (A = K, Rb, Cs) [43]; $Cs_5BiP_4Se_{12}$ [76]; $Cs_2CuP_3S_9$ [77]; $CuInP_2S_6$ [78]; $Zn_xBa_2B_2S_{5+x}$ (x ≈ 0.2) [44]
	Metal pnictidohalides and chalcohalides	$(Hg_6P_3)(In_2Cl_9)$ and $(Hg_8As_4)(Bi_3Cl_{13})$ [45], $(K_3I)[SmB_{12}(GaS_4)_3]$ [79], $[Sb_7S_8Br_2](AlCl_4)_3$ [80], $LaCa_2GeS_4Cl_3$ [81]

2 Classical Binary and Ternary Metal Pnictides and Chalcogenides

Generally, compound semiconductors are promising materials for mid-IR NLO devices because of their many advantages such as broad transparency range, large optical nonlinearity and large birefringence which make their phase-matching conditions exist for a variety of NLO interactions within the medium. (II, III)–VI and III–V semiconductors and II–IV–V_2 and I–III–VI_2 chalcopyrites are now the main applied mid-IR NLO crystals in commerce and laboratory, for example, a novel and promising terahertz source based on collinear phase-matched difference frequency generation (DFG) has been achieved in a GaSe crystal [82]. A 4.5–10.7-mm broadband mid-IR continuum was produced by optical parametric generation in orientation-patterned GaAs [83], and $CdSiP_2$ is employed in a nanosecond, 90°-phase-matched, singly resonant optical parametric oscillator pumped at 1,064 nm to produce idler pulses near 6.2 μm with an energy as high as 470 μJ at 10 Hz [84].

Although such type of crystals posses many mid-IR NLO advantages, many of them also have some shortcomings. Recently, mainly two approaches were developed to avoid the drawbacks in specific applications including the manufacturing of QPM orientation-patterned structures with highly nonlinear but isotropic semiconductors and the doping or mixing of nonlinear crystals to produce more complex ternary compounds or quaternary compounds with excellent NLO properties. In order to meet the need of laser device fabrication, some improving growth techniques are adopted to grow high-purity and high-quality large crystals of them.

2.1 *Binary (II, III)–VI and III–V Compounds*

Among the binary (II, III)–VI and III–V semiconductor mid-IR NLO crystals, like GaSe, CdSe, HgS, and GaAs, only GaSe and GaAs have attracted steady attention from many researchers and developers.

GaSe [85] is one of the most promising materials for conversion of optical frequencies in the infrared and terahertz spectral regions due to its extreme transparency range (0.65–18 μm), high SHG efficiency (d_{22} (10.6 μm) = 54 pm/V), large birefringence ($\Delta n \sim 0.35$ at 1 μm), and high thermal conductivity of 16.2 W/mK [36]. The crystal structure of GaSe is formed by a system of hexagonal layers of Ga and Se atoms. One atomic layer consists of two monatomic sheets of Ga sandwiched between two monatomic sheets of Se. The strong intralayer covalent interaction and weak van der Waals forces between the layers make GaSe a quasi-two-dimensional, highly anisotropic material, but highly cleavable along the planes parallel to the atomic layers and notoriously difficult to cut and polish, which is a fundamental requisite of any practical NLO material.

Doping with Group III and Group VI elements [86–88], both in the cation and anion sublattices, and even doping with $AgGaS_2$ and $AgGaSe_2$ [89], were applied to improve the optical and mechanical properties of GaSe, but degradation of some other properties may also exist after doping.

The GaSe crystals doped with indium show a strengthening of the mechanical properties and can be cut and polished in crystallographic direction selected according to phase-matching conditions for interacting light waves and possess better nonlinear properties when compared with pure GaSe [90]. The microhardness and optical absorption do not vary linearly with indium concentration but exhibit an overall increase with increasing indium with a minimum absorption coefficient at In content as 0.5 wt% [91]. The optical quality of GaSe crystals is rapidly degraded on In doping, and GaSe:In crystals become non-useful in nonlinear devices at In content of >3 wt%; an optimal doping level for the GaSe:In crystals is estimated to be between 0.5 and 1 wt% of In content. Variations of In doping over the crystal volume do not disturb phase-matching conditions, and GaSe:In crystals show low sensitivity to phase-matching angle tuning with temperature [92]. Doping with aluminum (2 wt %) results in its highest hardness, but even at relatively low doping levels, a degradation in optical quality of the crystal is found, and the long-wavelength cutoff decreases by 1 μm because of the incorporation of light element [93]. The tellurium doping weakly influences the mechanical properties but shifts transmission range to longer wavelength side and increases the absorption coefficient. In the case of doping with erbium, the SHG efficiency of 0.5 at% erbium-doped GaSe crystal was found to be 55.3 pm/V or 24% higher than that of a pure GaSe crystal. The improvement of SHG efficiency is attributed to the substitutive and interstitial doping of Er ion in GaSe unit cell [94].

With sulfur doping (0.01 ~ 3 wt%), the transparency range shifted toward shorter wavelength side by 0.02 μm, and micro hardness increases up to 17 kg/mm^2, and it is found that nonzero effective nonlinearity of GaSe crystal doped with sulfur is 1.5 times better than indium-doped GaSe crystal [93], and SHG efficiency in $GaSe_{0.91}S_{0.09}$ is 2.4 times of pure GaSe [95].

For the compounds doping, the GaSe:S crystal grown from the melt GaSe: $AgGaS_2$ (10.6 wt%) is identified as the GaSe:S (2 wt%) crystal with almost the same optical quality compared to that in GaSe:S (2 wt%) grown by the conventional technique. Both GaSe:S (2 wt%) crystals are characterized by the almost identical transparency curve and SHG phase-matching diagram that are shifted to short-wavelength range in comparison with pure GaSe. Eleven percent lower second-order nonlinear susceptibility coefficient d_{22} of these doped crystals was measured as compared to that in pure GaSe. Nevertheless, both of them possess higher efficiency in SHG than that in pure GaSe due to a higher optical quality and damage threshold, modified phase-matching conditions, and other physical properties. GaSe:S crystals possess the best set of physical properties for the frequency conversion within mid-IR among doped GaSe crystals at 2–3 wt% sulfur doping. In addition, GaSe:S (2 wt%) crystals grown from the melt GaSe:$AgGaS_2$ can be cut and polished at arbitrary direction because of its good hardness property, which can find many important applications for outdoor experiments [89]. For GaSe doped with the 10.4 wt% $AgGaSe_2$ compound, the nonlinear coefficient rises to the highest reported value of 75 pm/V among doped GaSe crystals [90].

It seems that the doping result of GaSe:S crystals is better than those of GaSe crystals doped with other elements, and GaSe:S crystals grown from compounds-doping melts may have better NLO properties than direct sulfur-doping method.

III–V-type compounds, especially GaAs, posses a face-centered cubic crystal lattice whose symmetry is characterized by the cubic crystal system, class -43*m*. GaAs has excellent potential as a mid-IR nonlinear optical material because of its large nonlinear susceptibility ($d_{36} = 170$ pm/V), broad transparency range (0.9–18 μm), low optical absorption, high thermal conductivity (52 W/mK), high laser-damage threshold, and well-developed material technology [36]. But it has a phase-matching problem, arising from the optical isotropy of its zinc-blende-type crystal structure, which can be solved by introducing a QPM structure. The QPM structure contains periodic inversion of the crystal orientation which is used to induce a periodic reversal of the sigh of the NLO coefficient. Because GaAs is not ferroelectric, a scheme, which is different from that for lithium niobate, using a combination of lithographic patterning and electric field poling to produce domains of alternating polarity [96], must be implemented to produce the patterned structures for QPM. So far, some efforts to fabricate QPM structures for GaAs have been reported [97–99].

Several quasi-phase-matched diffusion-bonded GaAs stacks with as many as 53 plates for frequency doubling of a CO_2 pulsed laser were fabricated, and SHG of a pulsed laser yielded up to 24% peak power efficiency at 20 MW/cm^2 for the longest stack [97]. Recently, An all-epitaxial technique was developed for preparing orientation-patterned GaAs templates by MBE growth of GaAs/Ge/GaAs, and lithography and etching were used to define the periodic pattern on a template [98, 99].

2.2 II–IV–V_2 and I–III–VI_2 Chalcopyrites

The II–IV–V_2 group like $ZnGeP_2$ and $CdSiP_2$ can be derived from the III–V compounds like GaP by ordered substitution of elements from the periodic table which maintains the valence electron count per crystallographic site constant at four on the group III site. Closest to the II–IV–V_2 group are the I–III–VI_2 group like $AgGaS_2$ and $LiInS_2$ which derive from the II–VI zinc-blende compounds like CdS in exactly the same way as the II–IV–V_2 compounds derive from the III–V compounds. This ordering results in not only a doubling of the unit cell along the *c*-axis and the structure becoming tetragonal from cubic but also the properties of the materials becoming different along directions parallel and perpendicular to the *c*-axis and leading to their birefringence. Some II–IV–V_2 and I–III–VI_2 chalcopyrite compounds show very large SHG efficiency, which has been suggested as being associated with the distortion of the two types of metal tetrahedral coordinations in the compounds [100–103].

Many of II–IV–V_2 and I–III–VI_2 chalcopyrites have been found to be promising mid-IR NLO materials in the past due to their unique properties like large SHG efficiency, wide transparency region, and large enough birefringence to permit phase matching. Although some of them are available commercially or quasi-commercially, some drawbacks of them must be overcome before wide application. For example, $AgGaS_2$ and $AgGaSe_2$ have low residual absorptions but poor thermal conductivities and anisotropic thermal expansions with different signs [36]; $ZnGeP_2$ has excellent nonlinearity and thermal conductivity, but multiphonon and residual absorption limit its transparency from both sides so that pump wavelengths should lie above 2 μm which corresponds to less than one-third of its bandgap [104]; $CdGeAs_2$ possesses extremely high nonlinearity but also exhibits absorption features, and low temperatures are required to avoid the residual losses [36]. Recently, some achievements of strengthening their properties and overcoming some drawbacks have been made, and some large crystals of them have been prepared for practical applications using improved growth methods.

As one of the most famous II–IV–V_2 chalcopyrites, zinc germanium phosphide, $ZnGeP_2$ (ZGP), is the mid-IR NLO crystal of choice for laser frequency conversion in the 2–8-μm spectral range by virtue of its high SHG efficiency ($d_{14} = 75$ pm/V) and thermal conductivity (35 W/mK) as well as reductions in near-infrared absorption achieved in crystals grown by the horizontal gradient freeze technique (Fig. 1). Recently, the growth of high optical quality, single-crystal ZGP boule with dimensions of $27 \times 39 \times 140$ mm^3 has been demonstrated (Fig. 2a) [105], and a low-loss (1 0 0)-oriented ZGP sample measuring $30 \times 30 \times 22$ mm^3 was fabricated. By adjusting the seeding orientation in single-crystal boules with dimensions of $17 \times 30 \times 140$ mm^3, low-loss ZGP optical parametric oscillator (OPO) samples of $20 \times 20 \times 16$ mm^3 have been produced. Besides allowing for larger aperture samples, the larger ZGP single crystals have the added benefit of lower absorption at both 2 μm ($\alpha_{2\mu m} < 0.05$ cm^{-1}) and 1 μm ($\alpha_{1\mu m}$ ~1.0 cm^{-1}) than the smaller scale crystals. Improvements to the fabrication and polishing of ZGP samples resulted in an increased laser-induced damage threshold (LIDT). The LIDT of antireflection-coated ZGP samples at 2.05-μm and 10-kHz pulse rate frequency was increased to 2 J/cm^2. The combination of increased aperture, lower absorption, and improved LIDT of ZGP has it suitable for high-energy applications.

Large, high-quality, optically transparent crystals of $CdSiP_2$ have been grown from a stoichiometric melt and also by the horizontal gradient freeze technique. Crack- and twin-free single crystals of sizes up to $70 \times 25 \times 8$ mm^3 have been produced (Fig. 2b) [106]. It is transparent out to 9 μm, and its refractive index is very close to 3.0 throughout most of its transparency range. Its birefringence of ~0.05 is large enough for phase matching 1-, 1.5-, and 2-μm lasers into the mid-infrared, and it has high thermal conductivity (13.6 W/mK) which is 10–15 times higher than existing mid-IR NLO crystals $AgGaS_2$ (1.4 W/mK) and $AgGaSe_2$ (1.0 W/mK). These properties, along with a relatively high melting point (1,133°C) make it attractive for high-average-power applications.

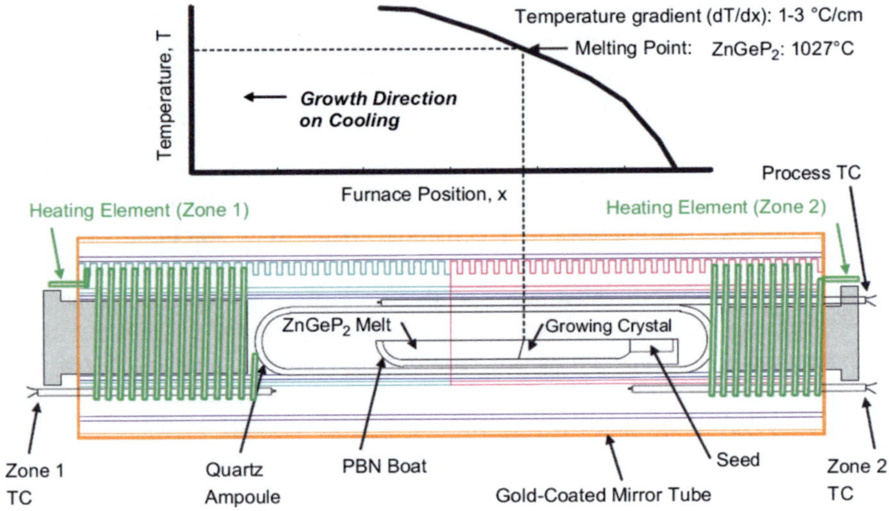

Fig. 1 Schematic of the transparent horizontal gradient freeze furnace used to grow $ZnGeP_2$. Adapted with permission from [105]. Copyright © 2008, Elsevier

$CdGeAs_2$ has outstanding nonlinear optical coefficients ($d_{14} = 236$ pm/V), broad infrared transparency range (2.3–17 μm), a sufficient birefringence (n_e–n_o ≈ 0.09) for phase-matched frequency generation, and a high thermal conductivity (about 4.18–6.69 $wm^{-1}\ K^{-1}$) for high-average-power application. These unique properties make $CdGeAs_2$ crystal a perfect candidate for SHG using CO_2 laser. However, some bad factors which influence the yield of high-quality $CdGeAs_2$ exist, such as huge thermal expansion anisotropy which makes the as-grown crystal extremely crack at the step of cooling and the synthesis processes often produce minor impurities like CdAs, $GeAs_2$, Cd_3As_2, etc. Recently, a large, crack-free $CdGeAs_2$ single crystal measuring 15 mm in diameter and 45 mm in length was grown in a vertical three-zone tubular furnace by a modified vertical Bridgman method (Fig. 2c) [107], i.e., quasi-seed technique with small temperature gradient and descending quartz ampoule, which can overcome the shortcomings met in the conventional vertical Bridgman method, such as cracking and polycrystallinity. High-purity, single-phase $CdGeAs_2$ polycrystallite for crystal growth was synthesized using a rocking furnace with temperature oscillation techniques. It is found that the cleavage plane of the as-grown crystal is (1 0 1) face, the crystal is integrated in structure and crystallized well, etch pits in the shape of pentagon on (1 1 2) face have been observed for the first time. And it is encouragingly indicated that the modified vertical Bridgman method is a convenient and effective way for high-quality $CdGeAs_2$ crystal growth. Other methods, such as the horizontal gradient freeze method together with a low thermal gradient [111], were also used to grow large crack-free $CdGeAs_2$ single crystals.

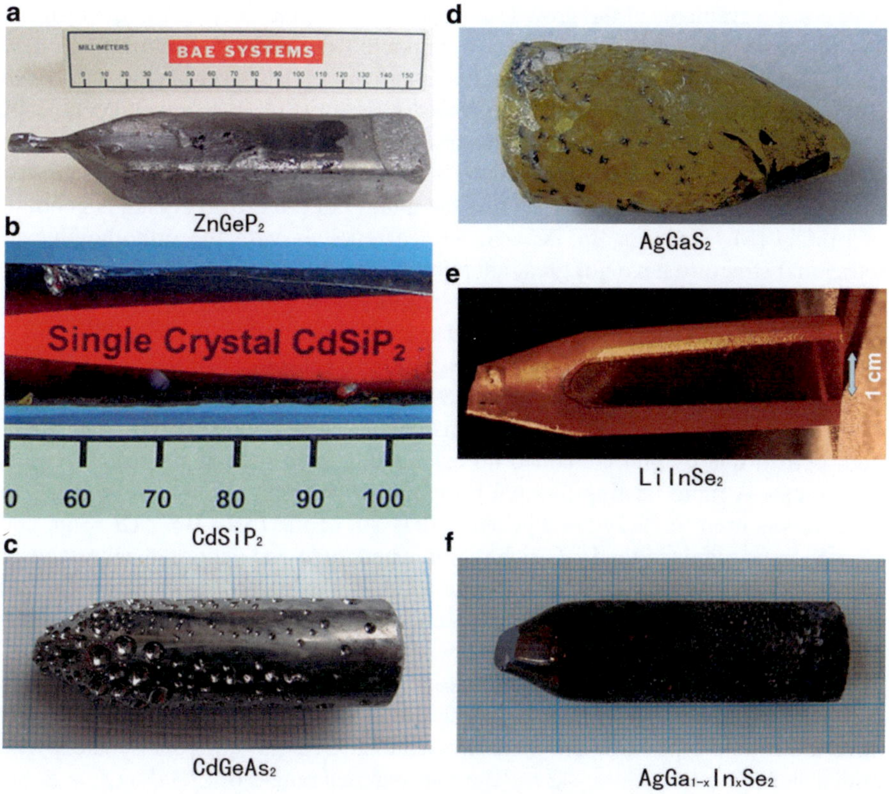

Fig. 2 Photograph of as-grown $ZnGeP_2$ (**a**), $CdSiP_2$ (**b**), $CdGeAs_2$ (**c**), $AgGaS_2$ (**d**), $LiInSe_2$, (**e**) and $AgGa_{1-x}In_xSe_2$ (**f**) crystals. Part (**a**) adapted with permission from [105]. Copyright © 2008 Elsevier. Part (**b**) adapted with permission from [106]. Copyright © 2010 Elsevier. Part (**c**) adapted with permission from [107]. Copyright © 2011 Elsevier. Part (**d**) adapted with permission from [108]. Copyright © 2006 Elsevier. Part (**e**) adapted with permission from [109]. Copyright © 2010 Optical Society of America. Part (**f**) adapted with permission from [110]. Copyright © 2009 Elsevier

In addition to II–IV–V_2 mid-IR NLO chalcopyrites like $ZnGeP_2$, $CdSiP_2$, and $CdGeAs_2$, which have received much attention recently, many of I–III–VI_2 chalcopyrites, especially $AgGaX_2$ (X = S, Se, Te) [100, 101, 112–115] and $LiMX_2$ (M = Al, Ga, In; X = S, Se, Te) [102–104, 116–118] families, also show very high promising mid-IR NLO properties, and much achievements have been made on growing large their high-quality big crystals.

Integral and crack-free $AgGaS_2$ single crystals with fewer defects were grown successfully and reproducibly in the a furnace based on a conventional vertical resistance-wound tubular two-temperature-zone furnace, which is modified by dividing the heating coils of the upper zone into two separate portions and adding a thermal conductor and a thermal baffle. During the large crystal growth, temperature gradients at the growth interface were kept at 30°C/cm in the furnace, and the descending rates of the growth ampoule were about 5–8 mm/day. A new cleavage face (1 0 1) was observed in the as-grown $AgGaS_2$ crystal (Fig. 2d) [108]. It was found that multiple diffraction peaks of the {1 0 1} faces are evident. Its IR transmittance is above 60%, and the absorption coefficient is 0.66 cm^{-1}. The results show not only that the modified furnace is a new promising furnace for the growth of $AgGaS_2$ single crystals but also that the quality of the grown crystal is good.

$LiMX_2$ (M = Al, Ga, In; X = S, Se, Te) crystals with the orthorhombic or tetragonal structure have got great attention due to relatively high SHG efficiency, wide bandgap, high enough thermal conductivity, low anisotropy of linear thermal expansion, and wide opportunities in phase-matching conditions [36, 104].

Bulk $LiMX_2$ crystals, including $LiInS_2$, $LiGaS_2$, $LiGaSe_2$, $LiInSe_2$, and $LiInTe_2$, were grown on oriented seeds by the BS technique in a vertical setup [119]. The crystals ~30 mm in length and ~15 mm in diameter were obtained [120]; they are of high optical quality and contained no extended defects such as inclusions, twins, and cracks. A rapid heating thermal technique with high pressures of inert gas in camera was used to study the T–x phase diagram of the Li_2C–B_2C_3 (B = In, Ga; C = S, Se) systems in the Li_2C-rich region where melt and vapor are aggressive to any container material [121]. A bounded solubility of two components in a liquid state is likely due to thermal dissociation of the melt into Li_2C, and $Li_{1+x}BC_2$ ($x < 0.1$) components were found.

Single crystals of $LiGaS_2$ and $LiGaSe_2$, up to 15 mm in diameter and 40 mm in length, of sufficient optical quality were grown using the BS technique [122]. They have a wurtzite-type structure $mm2$ whereas $LiGaTe_2$ is tetragonal chalcopyrite lattice with symmetry class –42 m. The transparency ranges of $LiGaX_2$ (X = S, Se, Te) are 0.32–11.6 μm, 0.37–13.2 μm, and 0.54–14.2 μm, respectively, at the absorption level of 5 cm^{-1}. The three refractive indices were measured in the whole transparency ranges of $LiGaS_2$ and $LiGaSe_2$, and n_a and n_c were found to be very close (quasi-uniaxial optical anisotropy) with a cross-point at 6.5 μm ($LiGaS_2$) and 8 μm ($LiGaSe_2$). Sellmeier equations were fitted, and phase-matching conditions for SHG were calculated: the 1.467–11.72-μm spectral range for the fundamental is covered by $LiGaS_2$ and $LiGaSe_2$.

$LiGaTe_2$ crystals have been grown by the BS technique [123, 124]. The clear transparency range of $LiGaTe_2$ extends from 2.5 to 12 μm, and its bandgap is 2.41 eV (515 nm) at room temperature. $LiGaTe_2$ is a positive uniaxial crystal which possesses sufficient birefringence for phase matching. Its nonlinear coefficient d_{36} estimated by phase-matched SHG is 43 pm/V ± 10% at 4.5 μm.

The main advantages of $LiInS_2$ are optical transparency over a wide range of wavelengths from 0.4 to 12 μm and high nonlinear susceptibility as $d_{33} = 18$ pm/V. The bulk $LiInS_2$ crystals for nonlinear optics were grown using the BS technique [125, 126]. Related to $LiInS_2$, $LiInSe_2$ is a nonlinear biaxial crystal and transparent from 0.54 to 10 μm at the 50% level (10-mm thickness). $LiInSe_2$ has been successfully grown in large size of 30 mm in diameter and 100 mm in length and with good optical quality (Fig. 2e) [109, 127]. With respect to $AgGaS(e)_2$ ternary chalcopyrite materials, $LiInSe_2$ displays a nearly isotropic thermal expansion behavior with three- to five-times-larger thermal conductivities associated with high optical damage thresholds, and low intensity-dependent absorption, allowing direct high-power downconversion from the near-infrared, especially 1,064 nm, to the deep mid-infrared. $LiInSe_2$ exhibits effective nonlinearity higher than that of the related $LiInS_2$ and comparable to that of $AgGaS_2$. Taking into account its similar transparency range, $LiInSe_2$ seems to be a good candidate for 1.064 μm pumped mid-IR OPO operation with the advantages of higher thermal conductivity and damage threshold.

Birefringence tailoring by means of mixing ternary chalcopyrite semiconductors [128, 129] offers a unique advantage for achieving noncritical phase-matched parametric interactions in the mid-IR, which are otherwise not accessible by periodically poled ferroelectric oxide crystals. Recently, many mixed chalcopyrite crystals have been grown for NLO applications.

Good quality of In-doping $AgGaSe_2$, $AgGa_{1-x}In_xSe_2$ ($x = 0.1$–0.3), single crystals with 20-mm diameter and 60-mm length have been grown by an improved Bridgman method (Fig. 2f), i.e., descending ampoule with rotation and the real-time temperature compensation technique [110]. The melt temperature oscillation method was used to synthesize high-purity single-phase $AgGa_{1-x}In_xSe_2$ polycrystal that is free of voids. The crystallization process was carried out in the two-layer quartz ampoule with a seed pocket. The temperature gradient at the growth interface in the furnace was normally 20°C/cm in the open bore, while it was 15°C/cm with the ampoule in the furnace. The transmittance of the crystal sample of 2-mm thickness is up to 65% in the region of 700–6,500 cm^{-1}. The results demonstrated that the improved new growth method of $AgGa_{1-x}In_xSe_2$ single crystals is promising and the quality of the grown crystals is good, and after annealing, the crystals can be used for IR NLO devices.

$AgGa_{1-x}In_xS_2$ with $x = 0.14 \pm 0.01$[130, 131] was found to be 90° phase matchable for the SHG of CO_2 laser radiation at 10.591 μm at 203°C. In addition, temperature-tuned 90° phase-matched DFG at 4.02 μm was demonstrated by mixing the idler output of a Nd:YAG third-harmonic pumped β-BBO optical parametric oscillator and its fundamental source at 1.0642 μm. $AgGa_{1-x}In_xSe_2$ crystal composed of 52.6% AgGaSe2 and 47.4% AgInSe2 is 90° phase matchable for temperature-tuned type I sum-frequency generation between the fundamental and the second-harmonic radiation of a CO_2 laser at 9.2714–10.5910 μm [132]. Temperature-tuned type II noncritical phase-matched SHG of CO_2 laser radiation at 9.27124–10.5910 μm was reported in $CdGe(P_xAs_{1-x})_2$ [133]. $AgIn_{1-x}Ga_xSe_2$ crystals grown by a vertical gradient freeze method [134] and by BS method [135] were also reported.

Phase-matching conditions are studied for SHG and OPO in $LiGa(S_{1-x}Se_x)_2$ [136] and $LiIn(S_{1-x}Se_x)_2$ [137] mixed crystals. Comparative analysis of their essential physical characteristics shows that the compounds seem to be among the most effective crystals for the creation of middle IR optical parametric oscillators and difference frequency generators under pumping by all solid-state and common near-IR lasers, including Nd:YAG laser. It seems that the most promising application of such crystals is the design of tunable mid-infrared femtosecond sources with one-stage frequency conversion of available near-IR femtosecond lasers.

2.3 Other Ternary Metal Chalcogenides

In addition to binary (II, III)–VI and III–V semiconductors, I–IV–V_2 and I–III–VI_2 chalcopyrites, much attention has been given to other ternary metal chalcogenides in the same category. NCS compounds $BaGa_4S_7$ [138] and $BaGa_4Se_7$ [139] were found to be new promising NLO materials in mid-IR region. They both exhibit phase-matching capability to cover the mid-IR spectral range by downconversion of 1,064-nm laser radiation. $BaGa_4S_7$ crystallizes in space group $Pmn2_1$ of the orthorhombic system, while $BaGa_4Se_7$ crystallizes in space group Pc of the monoclinic system. In the crystal structure of $BaGa_4Se_7$, $GaSe_4$ tetrahedra are connected to each other by corner sharing to form a three-dimensional framework with Ba cation in the cavities. Ba is in a "tighter" coordination environment of 8 Se compared with that of 12 S in $BaGa_4S_7$.

$BaGa_4S_7$ (Fig. 3a) [47, 48] crystal and its selenide analogue, $BaGa_4Se_7$ (Fig. 3b), [48] have been grown by a BS technique; linear and NLO properties were measured, indicating that $BaGa_4S_7$ has a good optical transparency region (545 nm to 9.4 μm) at an absorption level of 0.3 cm^{-1} from unpolarized transmission spectra, high second-order susceptibility coefficients (d_{33} = 12.6 pm/V), and high laser-damage threshold (1.2 J/cm^2 at 1.064 μm and a 15-ns pulse width). Transparency region of $BaGa_4Se_7$ is 0.776–14.72 μm.

$HgGa_2S_4$ [49, 142] is a negative uniaxial crystal having tetragonal I-4 space group and -4 point group symmetry; it can be viewed as a "defect" chalcopyrite derived from the I–III–VI_2 chalcopyrite structure by the order substitution of group II atoms and vacancies on the group I sites. $HgGa_2S_4$ possesses high nonlinear optical coefficients (d_{36} = 34 pm/V), wide transparency range of 0.5–13 μm, and appropriate birefringence ~0.06, and radiation resistance of $HgGa_2S_4$ is among the best ones for non-oxide NLO crystals. These promising properties make it useful for mid-IR NLO applications. $HgGa_2S_4$ has gotten much attention recently, not only on growing large crystals but also on device preparation [143–145]. $HgGa_2S_4$, which is at present in the development stage, was proved to provide better performance for DFG in terms of conversion efficiency than the commercially available $AgGaS_2$ due to its higher nonlinearity. Besides, it could find broad applications not only in the high-power femtosecond laser technology but also for downconversion of high-repetition-rate low-power femtosecond laser sources as well as of continuous-wave laser radiation in the future [146].

Tunable femtosecond pulse generation in the mid-IR between 6.6 and 11 μm is demonstrated by optical parametric amplification in Cd-doping $HgGa_2S_4$, $Cd_{0:35}Hg_{0:65}Ga_2S_4$, [147] pumped at 820 nm by a 1-kHz repetition rate Ti:sapphire regenerative amplifier. Femtosecond pulses at the idler wave with energies as high

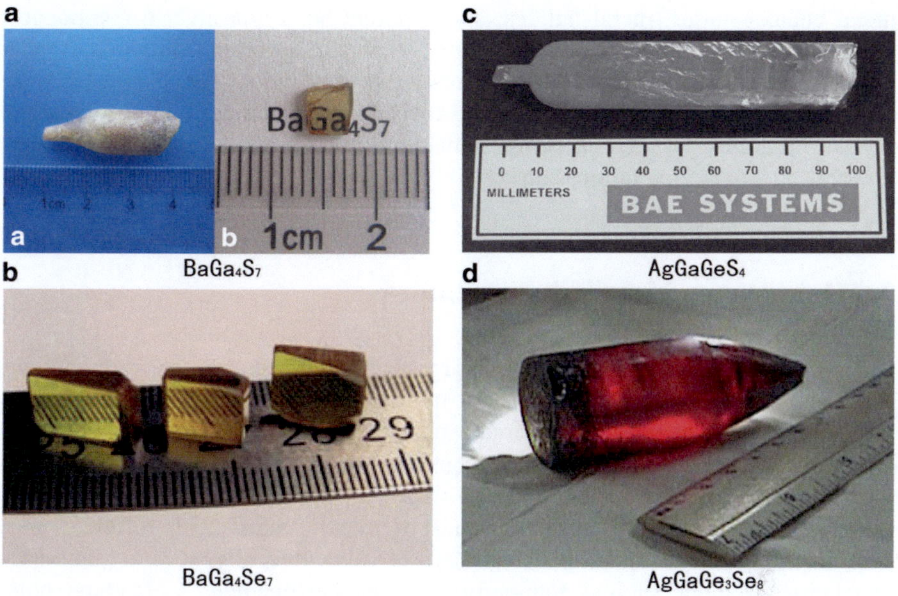

Fig. 3 photograph of as-grown or(and) polished $BaGa_4S_7$ (**a**), $BaGa_4Se_7$ (**b**), $AgGaGeS_4$ (**c**), and $AgGaGe_3Se_8$ (**d**) crystals. Part (**a**) adapted with permission from [47]. Copyright © 2009 American Chemical Society. Part (**b**) adapted with permission from [48]. Copyright © 2011 WILEY-VCH Verlag GmbH & Co. KGaA, Weinheim. Part (**c**) adapted with permission from [140]. Copyright © 2006 Elsevier. Part (**d**) adapted with permission from [141]. Copyright © 2010 Elsevier

as 2 μJ at 6.6 μm are produced by such a single conversion step from the Ti:sapphire spectral range. $Cd_xHg_{1-x}Ga_2S_4$ has the potential for NLO applications in the mid-IR, and its superior properties for frequency conversion of femtosecond pulses were experimentally demonstrated [50, 148].

New ternary β-$LaGaS_3$ has been synthesized from the stoichiometric mixture of elements by a conventional solid-state reaction at 1,100°C and annealed at 820°C [58]. The compound represents a new structure type that crystallizes in an NCS orthorhombic space group $Pna2_1$, and features the wavy GaS_4 tetrahedron chains that are separated by La^{3+} cations. Very weak SHG signal of the powders mostly ground from the handpicked β-$LaGaS_3$ crystals plus small amount of the as-synthesized powders was found.

In addition, two new ternary metal chalcogenides, α-Pb_2GeS_4 and Pb_2GeSe_4 [52], whose crystal structures contain isolated $[Ge(S, Se)]^{4-}$ tetrahedra and two different sites for the Pb^{2+} counter-cations, may produce large SHG responses in the mid- and far-IR region, although they show weak SHG effects under 1,064-nm laser due to absorption.

The development of classical binary and ternary metal pnictides and chalcogenides is more mature than other mid-IR NLO systems, such as quaternary metal chalcogenides, metal halides, and different-bond-type hybrid compounds, which will be reviewed in the rest chapters. Viewed as a whole, compared to other mid-IR NLO systems, they have two obviously advantages, good mid-IR NLO properties for practical applications and their simple chemical compositions and crystal structures, which can bring them more convenience during the material preparation and device fabrication.

3 Quaternary Metal Chalcogenides

Sulfur-based NLO materials have been largely used for frequency conversion in the IR region because of their improved IR transmission region (0.4–15 μm). Generally, compared with binary and ternary chalcogenides, quaternary chalcogenides are more difficult to be synthesized and grown into large crystals, but they are also attractive mid-IR NLO system due to a variety of acentric arrangements resulted from the combination of different kinds of metal centers with different size, coordination preference, and packing characteristics. Recently, several quaternary metal chalcogenides, such as diamond-like metal chalcogenides, which have practical using potential, were grown into large high-quality crystals, and many new mid-IR NLO materials, such as alkali-metal-based quaternary chalcogenides and other quaternary metal semiconductors, have been found in the system.

3.1 Quaternary Diamond-Like Metal Chalcogenides

Diamond-like metal chalcogenides mean that the structures of these compounds are derived from that of diamond, either the cubic or hexagonal form, and consist of tetrahedral building blocks [149, 150]. Compared with binary and ternary diamond-like phases, the compositional flexibility of quaternary materials increases and allows for the potential to tune the physical properties of the materials in nonlinear optics as well as other utilities.

$AgGaGe_nS(e)_{2(n+1)}$ ($n \neq 0$) is a very promising mid-IR NLO material system and may become an alternative to the widely used uniaxial $AgGaS_2$ due to their higher damage threshold and greater variety of phase-matching schemes [53, 151–154]. These quaternary compounds belong to *Fdd*2 "diamond" space group, and their symmetry class is orthorhombic *mm*2. The diamond-like structure of $AgGaGe_nS(e)_{2(n+1)}$ results from the substitution of Ge^{4+} by Ga^{3+} in the $GeS(e)_2$ cation sublattice. The valence deficiency is compensated by Ag^+ ions filling the tetrahedral vacancies.

The birefringence of quaternary semiconductors of the type $AgGaGe_nSe_{2(n+1)}$ ($n = 2, 3, 4, 5$), for example, $n_a–n_c$ at 1,064.2 nm, increases from 0.114 for $n = 2$ ($AgGaGe_2Se_6$) to 0.149 for $n = 5$ ($AgGaGe_5Se_{12}$), which are substantially larger than the birefringence of the uniaxial parent compound $AgGaSe_2$ (~0.022). All four quaternary compounds are optically negative biaxial crystals. Based on the type I interaction and propagation along the *Y* principal optical axis, the calculated SHG limit (minimum fundamental wavelength) is ~1,470 nm for $AgGaGe_2Se_6$ and ~1,240 nm for $AgGaGe_5Se_{12}$, which are much lower than the 3,120 nm for $AgGaSe_2$. Thus, the orthorhombic $AgGaGe_nSe_{2(n+1)}$ crystals can be used for SHG down to their band edge. On the average, the larger nonlinear coefficient d_{31} of $AgGaGe_nSe_{2(n+1)}$ ($n = 2, 3, 4$, and 5) is very close to d_{36} of $AgGaSe_2$ (~30 pm/V) while the d_{32} is roughly two times smaller [155]. Recently, $AgGaGeS_4$, $AgGaGe_5Se_{12}$, and $AgGaGe_3Se_8$ are the three representatives that have gotten more attention on growing large crystals in the $AgGaGe_nS(e)_{2(n+1)}$ system [140, 141, 156, 157].

$AgGaGeS_4$ [156] and $AgGaGe_5Se_{12}$ are promising NLO crystals for frequency-shifting 1-μm solid-state lasers into the mid-infrared (2–12 μm) spectral range. They were synthesized by vapor transport in sealed ampoules from high-purity elemental starting materials, and crystals were grown by the horizontal gradient freeze technique in transparent furnaces (Fig. 3c) [140]. $AgGaGe_5Se_{12}$ exhibited incongruent melting behavior, and small optical samples extracted from an as-grown polycrystalline boule had high scattering losses. $AgGaGeS_4$ growth was far more favorable, resulting in a crack-free single crystal measuring 19 mm in diameter and >80 mm in length with as-grown 2.05-μm absorption losses $<0.05\ cm^{-1}$. The measured laser-damage threshold of an uncoated $AgGaGeS_4$ crystal at 2.05 μm was 1.1 J/cm^2, and room-temperature measurements of thermal diffusivity, heat capacity, and thermal conductivity yielded values of 0.224 mm^2/s, 0.448 J/gK, and 0.399 W/mK, respectively. High-quality $AgGeGaS_4$ single crystal with 30-mm diameter and 80-mm length was also grown via direct reaction of raw materials $AgGaS_2$ and GeS_2, by using the modified BS spontaneous nucleation method [157]. The absorption coefficient in the region of 6.8–7.8 μm is as low as $0.02\ cm^{-1}$, and frequency doubling for 2.79 μm and 8 μm with different lasers was demonstrated successfully. The reaction conditions are easy to be maintained and controlled, which may provide a new method to produce other high-quality $Ag_xGa_xGe_{1-x}S_2$ materials via changing the amount of GeS_2.

Crystals of 19–25 mm in diameter and several centimeters long with $AgGaGe_5Se_{12}$ and $AgGaGe_3Se_8$ stoichiometric compositions were grown by the vertical Bridgman method (Fig. 3d) [141]. Optical measurements of parallel polished crystals show their transparency from 0.60 to 16 μm. Both crystals belong to the orthorhombic class and showed very good fabricability. The heat capacity of the $AgGaGe_3Se_8$ material was 0.28 J/gK, and the thermal conductivity was 0.30 W/mK. Experimental SHG gave the d_{31} coefficient of 29 pm/V for the both compositions. The OPO analysis indicates that type I phase matching is possible with the $AgGaGe_5Se_{12}$ crystal for conversion to the IR using 1.06-μm pump wavelength. The combination of transparency, low loss, and phase-matching results indicates that this material will enable 1.06-μm conversion to mid-IR and far-mid regions.

The substitution of Ag or Cu with Li widens the bandgap in diamond-like mid-IR NLO metal chalcogenides, increasing their potential in SHG applications

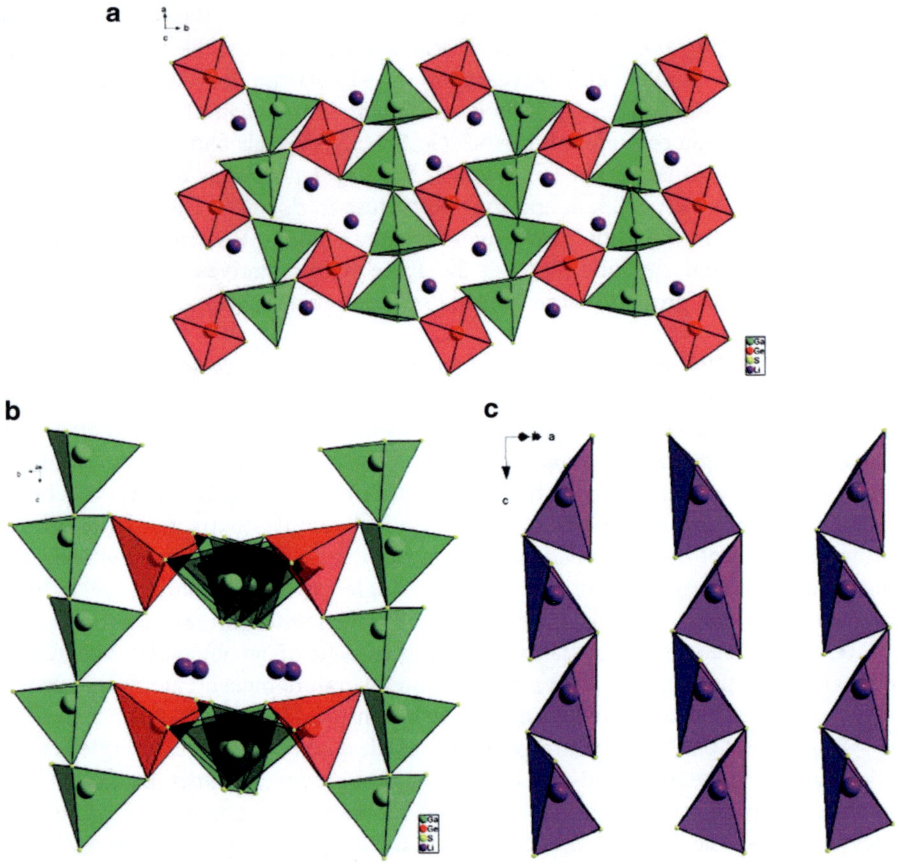

Fig. 4 (**a**) View of the $Li_2Ga_2GeS_6$ structure parallel to [001]. (**b**) View of the structure of $Li_2Ga_2GeS_6$ projected onto the (101) emphasizing GaS_4 chains. (**c**) View of the structure on to the (110) plane emphasizing the lithium tetrahedral chains. Adapted with permission from [54]. Copyright © 2008 American Chemical Society

because of the possibility of increased laser-damage thresholds. Additionally, the presence of Li as the univalent cation allows for the use of a lithium polysulfide flux as a reaction medium to promote crystal growth.

A new thiogermanium sulfide $Li_2Ga_2GeS_6$ has been synthesized [54], and its structure was found to be isomorphous with $AgGaGeS_4$, which is well known as a promising infrared NLO material. It crystallizes in the NCS orthorhombic space group *Fdd*2. The host structure is built of GaS_4 tetrahedra linked by corners to GeS_4 tetrahedra to create a 3-D framework forming tunnels along the *c*-axis (Fig. 4a, b). The Li atoms in distorted tetrahedral coordination form an infinite chain parallel to the *c*-axis and perpendicular to the *a*–*b* plane by corner sharing of the sulfur atoms of GaS_4 and GeS_4 tetrahedra (Fig. 4c).

The SHG efficiency determined on powders of $Li_2Ga_2GeS_6$ is ~200 times larger than that of α-SiO_2, and its transparent region is from ~0.4 to ~16.7 μm. Unlike $AgGaS_2$ and $AgGaGeS_4$, $Li_2Ga_2GeS_6$ was observed to be very stable under prolonged Nd:YAG 1.064-μm laser pumping, indicative of a large improvement in laser-damage threshold. This new material is very promising in the next generation of high-power infrared NLO applications.

The semiconductors Li_2CdGeS_4 and Li_2CdSnS_4, which are of interest for their nonlinear optical properties, were synthesized by using high-temperature solid-state and polychalcogenide flux reactions [55]. Both compounds were found to crystallize in $Pmn2_1$, and their structures are diamond-like with the tetrahedra orientating in the same direction along the *c*-axis. A modified Kurtz NLO powder technique was used to determine their SHG responses. Li_2CdGeS_4 displayed a type I phase-matchable response of approximately 70 × α-quartz, while Li_2CdSnS_4 displayed a type I non-phase-matchable response of approximately 100 × α-quartz. Diffuse-reflectance spectroscopy was used to determine bandgaps of 3.10 and 3.26 eV for Li_2CdGeS_4 and Li_2CdSnS_4, respectively. These properties as well as wide transparency region from ~0.4 to ~8.3 μm warrant further investigation of quaternary Li-containing diamond-like metal chalcogenides for NLO applications in the mid-IR region.

3.2 Alkali-Metal-Based Quaternary Chalcogenides

Alkali-metal-based quaternary chalcogenides in this chapter did not contain alkali-metal-containing diamond-like metal chalcogenides, which are reviewed in the previous section. Alkali-metal-based quaternary chalcogenides are a newfound family of mid-IR NLO materials in the past decade.

The closely related phases α- and β-$A_2Hg_3M_2S_8$ (A = K, Rb; M = Ge, Sn) [39] were discovered using the alkali polychalcogenide flux method; they present new structure types with a polar NCS crystallographic motif; both α-form with orthorhombic space group *Aba*2 and β-form with monoclinic space group *C*2 exhibit high environmental stability, optical transparency from ~500 nm to over 10 μm, and good NLO response, which is ten times larger than that of $LiNbO_3$. These materials (particularly β-$K_2Hg_3Ge_2S_8$) present good birefringence, good mechanical and thermal characteristics, and a high optical damage threshold. Its NLO coefficients exceed those of the commercially available $AgGaS_2$, and as a result, β-$K_2Hg_3Ge_2S_8$ promises to be suitable for a variety of long-wavelength NLO applications.

The crystal structure of α- and β-$A_2Hg_3M_2S_8$ (A = K, Rb; M = Ge, Sn) is similar and can be described in terms of a three-dimensional covalent anionic $(Hg_3M_2S_8)^{2-}$ (M = Ge, Sn) framework filled with K^+ or Rb^+ ions. Tetrahedral MS_4 (M = Ge, Sn) units and highly distorted seesaw-shaped HgS_4 tetrahedra share corner with each other to form layers of the stoichiometry $(Hg_2M_2S_8)^{4-}$ (M = Ge, Sn), which are then linked in the third direction via the linearly coordinated Hg

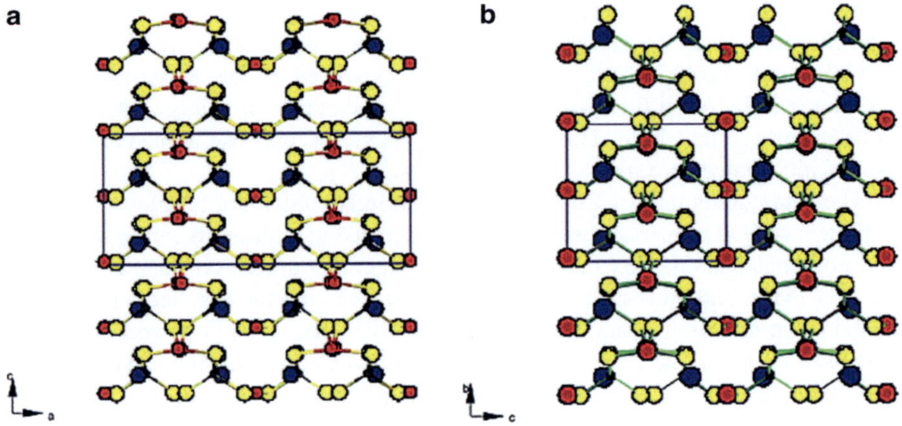

Fig. 5 Crystal structure of (**a**) α-$K_2Hg_3Ge_2S_8$, (s.g. *Aba*2). (**b**) Crystal structure of β-$K_2Hg_3Ge_2S_8$, (s.g. *C*2). *Blue circles* are K, *red circles* are Hg, *black circles* are Ge, and *yellow circles* are S atoms. Adapted with permission from [39]. Copyright © 2003 American Chemical Society

atoms to produce the polar three-dimensional framework. The K^+ or Rb^+ ions are surrounded by six $(MS_4)^{4-}$ (M = Ge, Sn) tetrahedra positioned in an octahedral fashion, and the sulfur atom cage defines the environment of K^+ and Rb^+ ions (Fig. 5).

Semiconductor $Na_{0.5}Pb_{1.75}GeS_4$ with bandgap of about 2 eV was synthesized [56] by reacting Pb and Ge in molten Na_2S_x at 530°C. The compound crystallizes in the cubic, NCS space group *I*-43*d*, which adopts a three-dimensional structure featuring $[GeS_4]^{4-}$ tetrahedral building blocks, which coordinate to sodium and lead cations. Preliminary experiments on powder sample of $Na_{0.5}Pb_{1.75}GeS_4$ using ~150 μJ laser light at 3.5 μm from the sample showed a SHG signal which was seven to eight times greater than that observed for similarly prepared powder samples of $LiNbO_3$, indicating that it may be a potential NLO material.

3.3 *Other Quaternary Metal Chalcogenides*

Four new quaternary isostructural rare-earth thiosilicates, $ZnY_6Si_2S_{14}$, $Al_{0.50}Dy_3(Si_{0.50}Al_{0.50})S_7$, $Al_{0.38}Dy_3(Si_{0.85}Al_{0.15})S_7$, and $Al_{0.33}Sm_3SiS_7$, crystallized in the chiral and polar space group $P6_3$, have been prepared by a facile synthetic routine with a new strategy using metal oxides in the air through the employment of a boron agent [40]. The first three compounds show strong SHG effects at 2.1 μm with the intensities of about 2, 2, and 1 times that of KTP, respectively. They are transparent from ~0.62 to ~14 μm. The four compounds are isostructural, and each of them contains three types of building units: AS_6 (A = Al or Zn) octahedra, MS_4 (M = Al or Si) tetrahedra, and LnS_8 (Ln = Y,

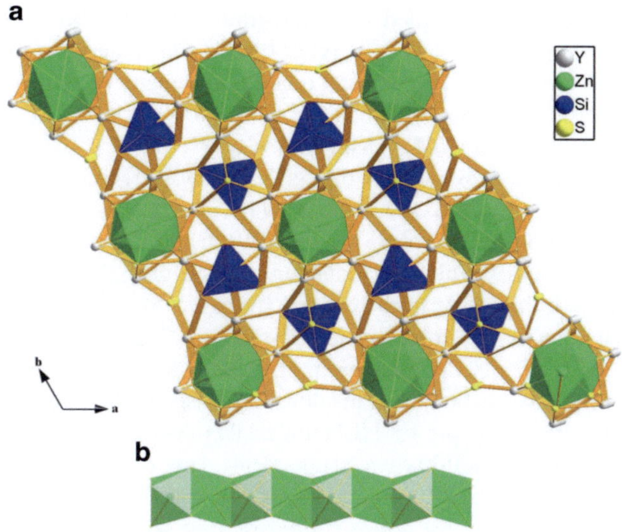

Fig. 6 View of $ZnY_6Si_2S_{14}$ along the *c* direction (**a**), *blue* tetrahedra are SiS_4 units, and the 1-D face-sharing chain of ZnS_6 octahedra (**b**, *green*)

Dy, Sm) bicapped trigonal prisms (BTPs). Their crystal structures can be described as a 3-D Ln-S open framework (Fig. 6a), of which the 1-D channels along the *c* direction and tetrahedral holes are occupied by AS_6 (A = Zn, Al) chains (Fig. 6b) and isolated MS_4 tetrahedra, respectively. Six LnS_8 BTPs connect to each other by sharing faces to accommodate octahedral interspaces for Zn^{2+} or Al^{3+} cations, and three LnS_8 BTPs create tetrahedral holes for M atoms by sharing corners.

$La_3CuGeSe_7$ [57] shows a weak SHG response (~30% α-SiO_2) when irradiated with 1,064-nm laser light. Its crystal structure contains isolated $(GeSe_4)^{4-}$ tetrahedra, trigonal planar $(CuSe_3)^{5-}$ units, and La^{3+} counter-cations.

4 Binary and Ternary Metal Halides

Hagemann and Weber published an important analysis on the structure of binary and ternary metal halides [158]. They discussed the physical, chemical, and crystallographic aspect and developed a concept for preparing NLO halides. Based on the bond-charge model, optical hyperpolarizabilities were calculated for more than 100 A–X bonds where X = Cl, Br, or I. Their tests showed that hyperpolarizabilities are a sound basis for estimating of NLO susceptibilities in halides. These materials have excellent damage resistance and high nonlinear figure of merit (higher than KDP and many other oxides). Among their attractive material properties are good mechanical properties (for ease of fabrication), reasonably high thermal conductivity coupled with transmittance over a broad spectral range (0.30–40 μm). Some practical mid-IR NLO binary and ternary metal halides have been grown into large crystals with high quality, and many new mid-IR NLO ones have also been found in this system, such as Tl_nMX_{n+2} (n = 3, 4; M = Pb, Hg; X = halogen) [61], $BaMF_4$ (M = Mg, Mn, Co, Ni, and Zn) [62, 159, 160], and ABX_3 (A = alkali metal; B = Ge, Cd, Hg; X = halogen) [63, 161–163].

$HgBr_2$ crystal of about 15 × 15 × 1.5 mm^3 in size was grown from ethanol solution by the lowering temperature method [59]. It shows a remarkable and phase-matchable powder SHG, which is about ten times higher than that of KDP. It is transparent through the whole mid-IR range (from 2.5 to 25 μm). Its laser-induced damage threshold was preliminarily tested to be as high as about 0.3 GW/cm^2 resulted from its relatively large bandgap of about 3.3 eV. It is stable in air below 100°C, and its crystals can be grown in solutions. Based on these results, we believe that it is a promising candidate for NLO materials to be used in the mid-IR region.

SbF_3 is a typical and well-known halide [164]. Due to the existence of the lone pair of electrons on the Sb^{3+} cation and its NCS trigonal–pyramidal configuration, the SbF_3 group should exhibit nonzero microscopic second-order NLO property. Meanwhile, SbF_3 crystallizes orthogonally with the NCS space group *C*2*cm* and has a net polarization along the [1 0 0] direction. Therefore, it should show macroscopic second-order NLO property. On the other hand, fluoride normally possesses a large bandgap which will lead to relatively large laser-damage threshold and high stability. SbF_3 single crystal with the size of 10 × 3 × 2 mm^3 has been grown by slow solvent evaporation in aqueous solution at constant temperature (Fig. 7a) [60]. The intensity of SHG effect is about 5.8 times as large as that of KDP and is phase matchable. The SbF_3 compound is transparent in the range of 0.29–12 μm. Its bandgap is calculated to be about 4.3 eV.

Tl_3PbBr_5, Tl_4PbI_6, Tl_4HgI_6, and Tl_3PbI_5 single crystals of centimeter size with optical quality were grown in vertical geometry and with capillary seeding (Fig. 7b) [165]. Crystals of 3:1:5 stoichiometry showed orthorhombic symmetry, and crystals of 4:1:6 stoichiometry showed tetragonal symmetry. Crystals were grown by Bridgman method by using varying growth rates between 1 and 3–5 cm/day, temperature gradients ranging 20–30 K/cm, and cooling rates 5–10 K/h. The microstructures showed self-poling during the growth of 3:1:5 classes of materials. The observed width of lamella is approximately 20 mm for the growth rate of 2 cm/day.

Centimeter-sized single crystals of Tl_3PbCl_5 (Fig. 7c) [166] and Tl_3PbBr_5 (Fig. 7d) [167] were successfully grown using the BS method in a silica ampoule sealed under HCl gas and Ar atmosphere, respectively. Growth initiated in a thin capillary leads to a single domain after a few centimeters of growth, and the resulting single crystals, typically 8 cm long for Tl_3PbCl_5, are colorless with a hardly elliptically deformed circular cross section, while Tl_3PbBr_5 crystals are typically 1–2 cm long, yellow, and transparent. The both compounds are nonhygroscopic. The phase transition of Tl_3PbBr_5 undergone upon cooling at ~237°C is of first order with fast kinetics, and phase transition undergone by Tl_3PbCl_5 is at 171°C. The transparent ($T > 0.5$) regions of Tl_3PbCl_5 and Tl_3PbBr_5 are 0.5–20 μm and 0.65 to at least 24 μm, respectively.

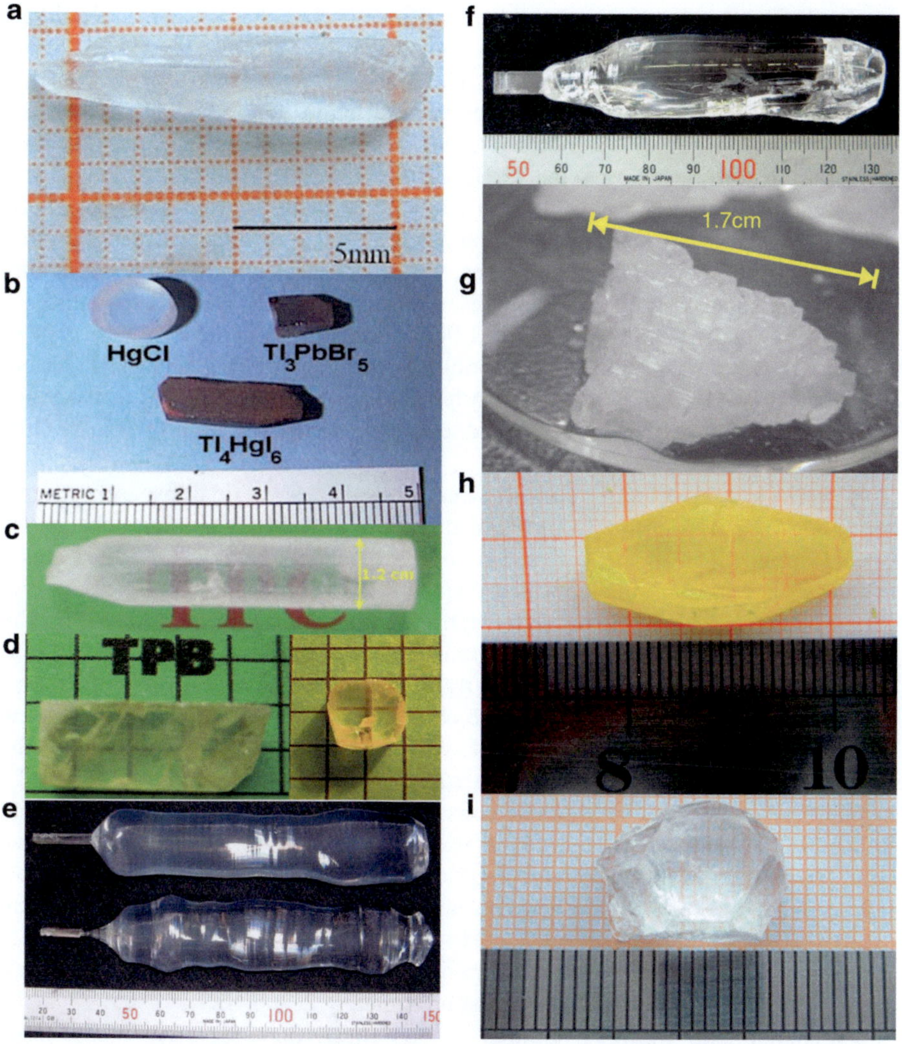

Fig. 7 photograph of as-grown SbF_3 (**a**), Hg_2Cl_2, Tl_3PbBr_5, Tl_4HgI_6 (**b**), Tl_3PbCl_5 (**c**), Tl_3PbBr_5 (**d**), $BaMgF_4$ (**e**), $SrAlF_5$ (**f**), $RbGeCl_3{\cdot}x(H_2O)$ (**g**), $Cs_2Hg_3I_8$ (**h**), and $NaSb_3F_{10}$ (**i**) crystals. Part (**a**) adapted with permission from [60]. Copyright © 2008 Elsevier. Part (**b**) adapted with permission from [165]. Copyright © 2005 Elsevier. Part (**c**) adapted with permission from [166]. Copyright © 2006 Elsevier. Part (**d**) adapted with permission from [167]. Copyright © 2006 Elsevier. Part (**e**) and (**f**) adapted with permission from [38]. Copyright © 2005 Elsevier. Part (**g**) adapted with permission from [168]. Copyright © 2009 Japan Society of Applied Physics. Part (**h**) adapted with permission from [65]. Copyright © 2008 American Chemical Society. Part (**i**) adapted with permission from [66]. Copyright © 2009 American Institute of Physics

The Tl_4HgI_6 crystals were grown by the BS method from a stoichiometric charge in a vertical furnace [169, 170]. It is shown that the Tl_4HgI_6 crystal melt congruently at the temperature of 396°C. The crystal grown from the stoichiometric charge has red color. When the iodine concentration in the charge exceeds the stoichiometric value, the color of crystal becomes black. In the vicinity of the melt composition corresponding to the Tl_4HgI_6 compound, Tl_2HgI_4 compound was also found with the melting temperature of 318°C. The black-phase Tl_4HgI_6 crystal is optically positive, transparent in the range from 1.2 to 40 μm, and has the point symmetry group C_{4v}. In OPO processes pumped by 1.32-μm radiation, the tuning range of the signal and idle wavelengths extends from 1.4 to 40 μm. Apparently, it is possible to realize OPO processes at 1.0642-μm pumping in the red-phase crystal. Measurements of the nonlinear susceptibility coefficient of high-quality (in particular, red phase) Tl_4HgI_6 crystal, which has the absorption coefficient in the transparency range below 0.05, will allow one to estimate its potential for use in OPO devices.

$BaMgF_4$ is a ferroelectric fluoride which shows a very wide transparency range extending from 125 nm to 13 μm, which confers to $BaMgF_4$ a unique chance for optical applications in the UV and mid-IR wavelength regions. $BaMgF_4$ single crystal with 1 in. in diameter was grown by the Czochralski technique (Fig. 7e) [38]. The crystal belongs to the pyroelectric fluoride family $BaMF_4$ (M = Mg, Co, Ni, Zn) with space group $Cmc2_1$ and transparency range from ~125 nm to 13 μm.

QPM can be achieved in the whole transparency region of $BaMgF_4$, in contrast to birefringent matching, which can be obtained in a limited range 573–5634 nm. First demonstration of SHG by QPM with a ferroelectric fluoride is shown by frequency doubling the emissions of a 1,064-nm Nd:YAG laser and a tunable Ti: sapphire laser [171]. The shortest emission is obtained in the UV at 368 nm, indicating the potential of $BaMgF_4$ as nonlinear medium for the fabrication of all solid-state lasers in the vacuum-UV/UV and mid-IR wavelength regions. At low temperature, high-yield hydrothermal route has been discovered for the phase-pure synthesis of the multiferroic $BaMF_4$ (M = Mg, Mn, Co, Ni, and Zn) family [172]. The powder SHG measurements determined that all of the materials as type I phase matchable with an SHG efficiency of approximately 2 × α-SiO_2.

Compared with $BaMgF_4$, another ternary metal fluoride, $SrAlF_5$ is a more promising for the realization of all solid-state lasers by the QPM technique due to its larger effective nonlinear coefficient [38]. $SrAlF_5$ has been suggested to belong to the class of uniaxial ferroelectric [173] and its large single crystals which were grown by the Czochralski technique (Fig. 7f) [38].

Recently, many ternary halides with perovskite structure, ABX_3 (A = alkali metal; B = Ge, Cd, Hg; X = halogen)-type inorganic ternary halides, became a new category of nonlinear optical materials with potential application from visible to infrared.

Single crystals of $RbGeCl_3{\cdot}x(H_2O)$ (HRGC), sized up to 3 × 2 × 1 cm^3, were grown in aqueous solution by a slow dehydrate technique (Fig. 7g) [168]. In comparison with the known NLO material KDP, HRGC's NLO susceptibility is about one-third of that for KDP. The absorption edge of HRGC occurred at 310 nm

(~4.0 eV), which indicates that the NLO HRGC crystal can have larger laser-damage threshold. According to the Fourier transform infrared measurement, HRGC has a transparent region from 0.31 to 30.84 μm; thus, it can be applied to wider optical spectrum from ultraviolet, visible, to mid-IR.

$CsGeCl_3$ adopts distorted perovskite structure with *R*3 space group. $GeCl_3^-$ has a similar structure to that of TiO_3^{2-} in the $BaTiO_3$ lattice; $GeCl_3^-$ anion group exists as $GeCl_6$ octahedron structure in the crystal lattice and has a distortion along the *three* axes [174, 175]. Transparent $CsGeCl_3$ crystal of about 5 × 5 × 5 mm^3 has been grown from a HCl–EtOH–CsCl mixture solution. Its transparency region was found to be up to 20 μm. It has no congruent melting point and decomposes at 325°C. Powder double-frequency studies show that its SHG signal was about five times to that of KDP and damage threshold was up to 200 MW/cm^2 [176].

According to the powder X-ray diffraction pattern and powder SHG results, an innovative infrared NLO crystal $CsGeBr_3$, which was characterized as a rhombohedral crystal structure, was synthesized. Ab initio calculations on $CsGeBr_3$ were also carried out to analyze the related electronic and optical properties. The space group symmetry of rhombohedral $CsGeBr_3$ was found to be *R*3*m*. The powder SHG measurement of $CsGeBr_3$ also showed that its NLO efficiency was larger than that of rhombohedral $CsGeCl_3$ by about 1.62 times and KDP by about 9.63 times, respectively [177] Saturated powder SHG integration results of increasing powder particle sizes revealed that rhombohedral $CsGeBr_3$ is phase matchable. Its infrared transparent spectrum is extended to more than 22.5 μm. And more importantly, its mixing halide $CsGe(Br_xCl_{1-x})_3$ [64, 178, 179] is phase matchable, and the infrared transparent spectrum of rhombohedral $CsGe(Br_xCl_{1-x})_3$ is extended to more than 30 μm.

$RbCdI_3{\cdot}H_2O$ single crystal with the size of 20 × 4 × 2 mm^3 has been grown by solvent slow evaporation in aqueous solution [180]. The intensity of SHG effect has been measured by using the Kurtz powder technique to be about 3.6 times as large as that of KDP. The $RbCdI_3{\cdot}H_2O$ crystal is transparent in most of IR region, and the transparent ranges are 2.5–2.7, 3.0–5.9, and 6.6–25 μm. Its bandgap is calculated to be 4.1 eV.

An NCS crystal structure of $CsCdBr_3$ has been successfully observed with X-ray single-crystal structure analysis. It crystallizes in the hexagonal space group $P6_3mc$. The structure contains $CdBr_6$ octahedra, which are connected in a plane-sharing way to form one-dimensional chain. Each octahedron is slightly distorted with three Cd–Br bond lengths being 2.774 Å and the other three Cd–Br bond lengths being 2.804 Å. The distortion directions of all the octahedra are almost parallel and give rise to the accumulation of the microcosmic NLO coefficient. The Kurtz powder technique shows that $CsCdBr_3$ has a powder SHG of about two times as large as that of KDP. It shows excellent transparency in the visible and infrared regions, and its thermal stability is also good [181].

$Cs_2Hg_3I_8$ single crystal with the size of 25 × 14 × 5 mm^3 has been grown by slow solvent evaporation in acetone at constant temperature (Fig. 7h) [65]. The intensity of SHG effect is similar to that of KTP, and the effect is phase matchable. The compound is transparent in the range of 0.5–25 μm. Its bandgap is calculated to be about 2.56 eV. Owing to these properties, $Cs_2Hg_3I_8$ appears to be a new potential NLO crystal applicable in the infrared region.

The NCS crystal $NaSb_3F_{10}$ exhibits a phase-matchable SHG effect 3.2 times as strong as KDP. It is thermally stable up to 220°C and transparent in the range of 0.25–7.8 μm, and its bandgap is about 5.0 eV. An experimental measurement indicates that its laser-induced damage threshold is as high as about 1.3 GW/cm^2, which is higher than all those IR NLO crystals reported in the literature. Moreover, $NaSb_3F_{10}$ single crystal with the size of 12 × 10 × 8 mm^3 has been grown by slow solvent evaporation in water at constant temperature (Fig. 7i) [66].

5 Different-Bond-Type Hybrid Compounds

The so-called different-bond-type hybrid compounds contain at least two obvious different types of bonds that have no absorption peaks in the mid-IR region. Unlike the mid-IR NLO compounds mentioned above, metal pnictides, chalcogenides, and halides, in which only metal–nonmetal bonds, i.e., metal– pnicogen, metal–chalcogen, and metal–halogen bonds, exist, respectively, metal thiotellurites, chalcopnictates, and thioborates simultaneously contain metal– nonmetal bonds, i.e., metal–chalcogen bonds and nonmetal–nonmetal bonds, i.e., Te–S, chalcogen-–chalcogen, and B–S bonds for the three types of compounds, respectively. Other systems, metal pnictidohalides and chalcohalides that contain metal–halogen and metal–pnicogen or metal–chalcogen bonds simultaneously, were also found to be mid-IR NLO material systems. Up to now, almost all preliminary experimental results of SHG efficiencies, optical transparency, and other NLO performances of the different-bond-type hybrid compounds are based on the powder samples; research work in this field focuses on exploring new potential mid-IR NLO crystals rather than growing larger ones.

5.1 *Metal Thiotellurites, Chalcopnictates, and Thioborates*

The thiotellurite unit is particularly polarizable because of the presence of a stereochemically active lone pair on the Te atom and the presumed flexibility of the S–Te–S angles. The layered compounds $RbAg_2TeS_6$ and $CsAg_2TeS_6$, formed from mixed Te/S polychalcogenide fluxes [42], crystallize in the NCS space group $P6_3cm$. The structures are composed of neutral $[Ag_2TeS_3]$ layers alternating with charge-balanced salt layers containing polysulfide chains of $[S_6]^{2-}$ and alkali-metal ions. $RbAg_2TeS_6$ and $CsAg_2TeS_6$ are air- and water-stable, wideband-gap semiconductors (e.g., ~2.0 eV) exhibiting NLO SHG of about 20% of that generated by $AgGaSe_2$. These materials also demonstrate phase-matchable SHG characteristics. The NLO properties are thought to arise from the NCS orientation of the polysulfide chains between the centrosymmetric layers of $[Ag_2TeS_3]$, indicating that even small subtleties in the structure can affect the functional properties of materials.

$LiAsS_2$ and mixed alkali-metal analogue $Li_{0.6}Na_{0.4}AsS_2$ crystallize in the space group *Cc*, their crystal structure feature polymeric $_\infty{}^1[AsS_2{}^-]$ chains, which are made from corner-sharing distorted AsS_3 trigonal pyramids. The alkali-metal ions are in a distorted-octahedral environment of sulfur atoms [70] (Fig. 8). $LiAsS_2$ and $Li_{0.6}Na_{0.4}AsS_2$ are direct-gap semiconductors with bandgaps of 1.60 and 1.75 eV, respectively. NLO SHG efficiency of the two compounds was measured with a modified Kurtz–NLO system using a 1,580-nm laser and found that $LiAsS_2$ is about 10 times and $Li_{0.6}Na_{0.4}AsS_2$ is about 30 times more efficient than $AgGaSe_2$. The highly polar structure that results from the alignment of dipoles in $LiAsS_2$ is responsible for the enhanced SHG intensity compared to weakly polar $AgGaSe_2$. This enhanced SHG intensity of the sodium-substituted $Li_{0.6}Na_{0.4}AsS_2$ could be attributed to the higher polarizability of sodium vis-à-vis lithium in the structure.

β-$LiAsSe_2$ crystallizes in the NCS space group *Cc* and is isostructural with $LiAsS_2$. It consists of densely packed $_\infty{}^1[AsSe^{2-}]$ polymeric chains and alkali-metal ions. The alignment of the chains parallels to the *c*-axis. The distorted $AsSe_3$ pyramid provides the molecular building units in the $_\infty{}^1[AsSe^{2-}]$ chains via corner bridging. The alkali-metal ions are in a distorted-octahedral environment of Se atoms, and the $LiSe_6$ octahedra form a three-dimensional network by itself through edge sharing in the *bc* plane and corner sharing along the *a* direction. γ-$NaAsSe_2$

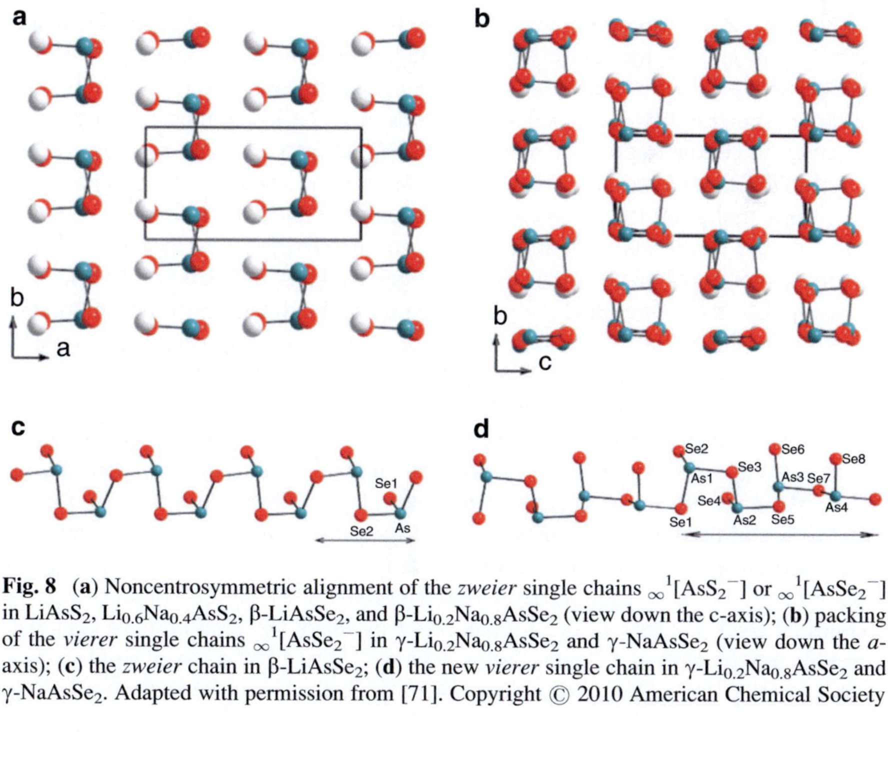

Fig. 8 (**a**) Noncentrosymmetric alignment of the *zweier* single chains $_\infty{}^1[AsS_2{}^-]$ or $_\infty{}^1[AsSe_2{}^-]$ in $LiAsS_2$, $Li_{0.6}Na_{0.4}AsS_2$, β-$LiAsSe_2$, and β-$Li_{0.2}Na_{0.8}AsSe_2$ (view down the c-axis); (**b**) packing of the *vierer* single chains $_\infty{}^1[AsSe_2{}^-]$ in γ-$Li_{0.2}Na_{0.8}AsSe_2$ and γ-$NaAsSe_2$ (view down the *a*-axis); (**c**) the *zweier* chain in β-$LiAsSe_2$; (**d**) the new *vierer* single chain in γ-$Li_{0.2}Na_{0.8}AsSe_2$ and γ-$NaAsSe_2$. Adapted with permission from [71]. Copyright © 2010 American Chemical Society

crystallizes in the NCS space group *Pc*. The structure also consists of densely packed parallel $_{\infty}{}^{1}[AsSe^{2-}]$ polymeric anionic chains and alkali-metal ions, but the conformation of the $_{\infty}{}^{1}[AsSe^{2-}]$ chain is different from that of β-polymorphs. The alkali-metal ions are in a distorted-octahedral environment of Se atoms. The edge sharing and corner sharing of the $NaSe_6$ octahedra give rise to a densely packed three-dimensional structure, which is different from the $LiSe_6$ packing in β-$LiAsSe_2$.

The NLO SHG response of two mixed alkali-metal analogues β-$Li_{0.2}Na_{0.8}AsSe_2$, γ-$Li_{0.2}Na_{0.8}AsSe_2$, and γ-$NaAsSe_2$ was measured with a modified Kurtz and Perry method over a range of wavelengths (1,000–2,000 nm) using a laser source. In the wavelength range of 700–900 nm, relative SHG efficiencies of β-$Li_{0.2}Na_{0.8}AsSe_2$, γ-$Li_{0.2}Na_{0.8}AsSe_2$, and γ-$NaAsSe_2$ were ~ 55, 65, and 75 times stronger than that of $AgGaSe_2$, respectively. Particle-size-dependent SHG measurements indicated that all three materials are type I non-phase matchable at 790 nm. The highly polar structure that results from the alignment of dipoles in the structure is likely responsible for the enhanced SHG intensity of these selenoarsenates.

An enormously valuable insight was shown that it is not the more polarizable nature of Na^+ over Li^+ that is responsible for the higher SHG efficiency of the Na-rich systems but the decrease in dimensionality that results from the larger Na^+ ion. As the Na^+ ions substitute for Li^+, the $_{\infty}{}^{1}[AsSe^{2-}]$ chains are increasingly pried apart from one another and interchain interactions diminish, thereby lowering the dimensionality of the system. Thus, controlling dimensionality should be a good strategy in designing better SHG materials.

The extremely strong nonlinear optical SHG response of $AAsQ_2$ species makes them very promising for future NLO investigations. The large polarity of $AAsQ_2$ chalcoarsenate materials that originates from the asymmetric pyramidal AsQ_3 units, the large polarizability of the selenium atoms, and the low dimensionality of the structures all couple to produce record-breaking SHG coefficients, as suggested by ab initio DFT theoretical investigations. The investigations of this system provide a new insight and point to a very valuable design principle for NLO materials [71].

Phase-pure syntheses of of $K_3Ta_2AsS_{11}$ and $Rb_3Ta_2AsS_{11}$ were achieved with the alkali-metal polythioarsenate flux method using A_2S/Ta/As/S (A = K, Rb) mixtures. The two compounds are isostructural and crystallize in the monoclinic space group *Cc*. Their structures consist of parallel $_{\infty}{}^{1}[Ta_2AsS_{11}{}^{3-}]$ polymeric anionic chain, which is a polysulfide species, building up of bimetallic $[Ta_2S_{11}]^{6-}$ units linked with AsS_3 pyramids, and K^+ or Rb^+ are embedded in the space between the chains. Electronic absorption spectroscopy of the solid samples showed a sharp absorption edge at 2.21 eV for the two compounds. In the range 700–900 nm, the SHG efficiency of the two compounds was ~15 times stronger than that of $AgGaSe_2$, and particle-size-dependent SHG measurements indicated that they are type I non-phase matchable at 770 nm. It was found that alkali-metal polarizability makes negligible contribution to the SHG efficiency of the compounds and the pyramidal $[AsS_3]^{3-}$ unit in the compounds may make the predominant [72]. The combination of two asymmetric units, $[Ta_2S_{11}]$ and $[AsS_3]$, in a single strand coupled with polar packing of strands appears to lead to strong NLO SHG response.

This implies that the approach of combing different asymmetric fragments (e.g., chalcogenides) to impart strong polarity in extended structures could be promising in finding exceptional candidate materials for NLO applications.

Polar structure $K_2P_2Se_6$ exhibits excellent mid-IR transparency from 0.6 to 19.8 μm and a strong SHG response. The SHG response is type I phase matchable and, in the wavelength range of 1,000–2,000 nm, was measured to be 50 times larger than that of the commercially used material $AgGaSe_2$, making it a potential contender material for applications in the IR region.

$K_2P_2Se_6$ crystallizes in the chiral trigonal space group $P3_121$ at room temperature. The structure has infinitely extended helical chains parallel to the *c*-axis. The chains are composed of ethane-like [P_2Se_6] repeat units that are linked via terminal Se–Se linkages to give infinite helices of $_\infty{}^1[P_2Se_6{}^{2-}]$ [73] (Fig. 9). The remarkably stronger response is attributed to the helical structure of the selenophosphate which is highly polar, in contrast to the weakly polar chalcopyrite structure of $AgGaSe_2$.

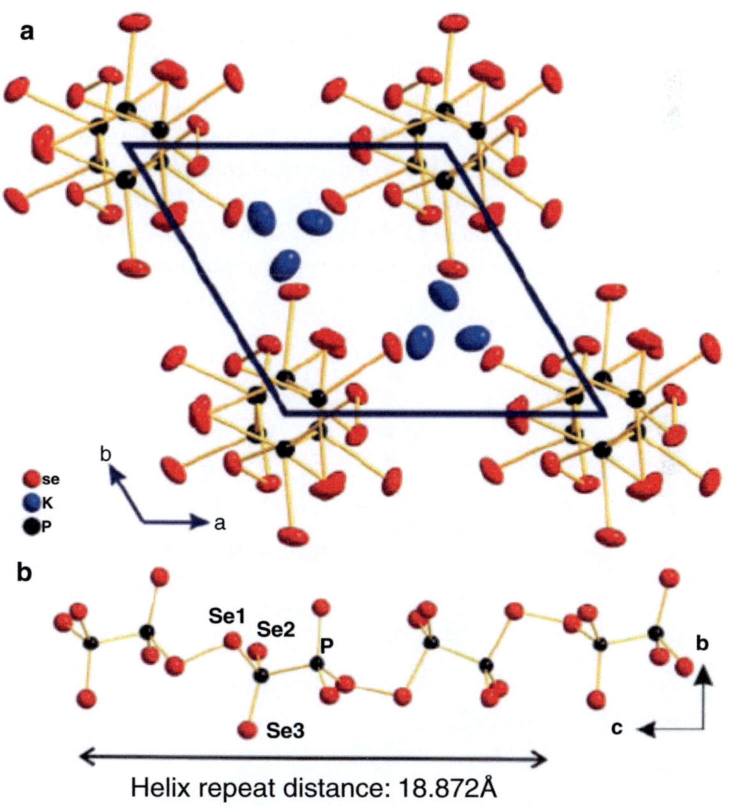

Fig. 9 Structure of $K_2P_2Se_6$ at 298(2) K. (**a**) The unit cell viewed down the *c*-axis. The thermal ellipsoids with 30% probability are shown. (**b**) View of a $_\infty{}^1[P_2Se_6{}^{2-}]$ chain looking down the *a*-axis. A helix forms by three [P_2Se_6] units and repeats itself at every 18.872 Å. Adapted with permission from [73]. Copyright © 2007 American Chemical Society

The glassy form of $K_2P_2Se_6$ also exhibited a surprising SHG response with no poling treatment.

One-dimensional polar selenophosphate compounds $APSe_6$ (A = K, Rb) [74, 182], which show crystal-glass phase-change behavior, exhibit strong SHG response in both crystal and glassy forms. The structure of the crystalline $APSe_6$ has infinite chains of $_\infty{}^1[PSe_6{}^-]$, consisting of the $[PSe_4]$ tetrahedral units condensed with Se_2 linkages, and A^+ counter-cations. The crystalline materials are type I phase matchable with very large SHG coefficients $\chi^{(2)}$ of 151.3 and 149.4 pm/V for K^+ and Rb^+ salts, respectively. The glass of $APSe_6$ exhibits comparable SHG intensities to the infrared NLO material $AgGaSe_2$ without any poling treatments, and $APSe_6$ exhibits excellent mid-IR transparency. Optical glass fibers with strong, intrinsic, and temporally stable second-order NLO response can be obtained, starting from $APSe_6$ (A = K, Rb). The as-prepared glass fibers exhibit SHG and DFG responses over a wide range of wavelengths. It is proposed that this approach can be widely applied to prepare permanent NLO glass from materials that undergo a phase-change process.

The new compound $Cs_5P_5Se_{12}$ [75] crystallizes in the NCS nonpolar space group *P*-4. It features the discrete molecular $[P_5Se_{12}]^{5-}$ anion with two types of formal charge, 3+ and 4+, on P. The trivalent formal charge is found on P(3) which is a central P atom chelated with two ethane-like $[P_2Se_6]^{4-}$ units to form a novel octahedral complex (Fig. 10). It has good optical transparency from the edge of the energy gap (~2.17 eV) to the mid-IR region. The SHG intensity of $Cs_5P_5Se_{12}$ was approximately equal to that of $LiNbO_3$ and 25% that of $AgGaSe_2$, and it is type I non-phase matchable. $Cs_5P_5Se_{12}$ exhibits phase-change behavior by forming

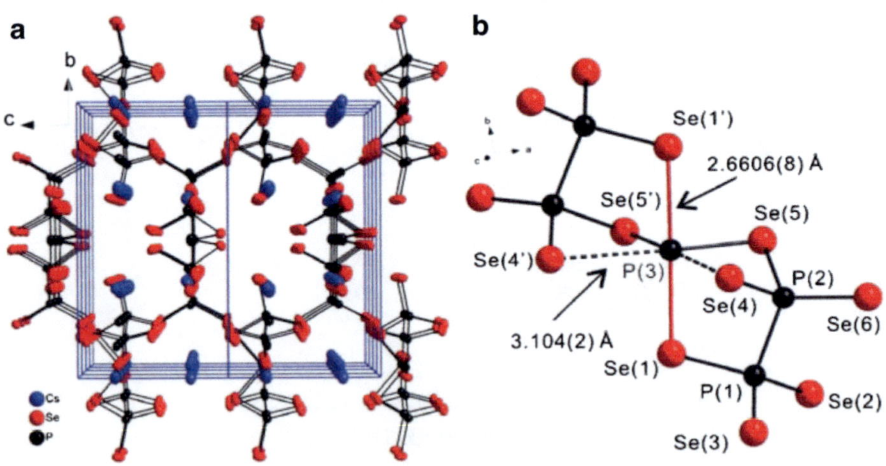

Fig. 10 (**a**) The noncentrosymmetric structure of $Cs_5P_5Se_{12}$. The thermal ellipsoids are shown with 60% probability. (**b**) $[P_5Se_{12}]^{5-}$ anion. *Red solid line* denotes long P–Se bonding at P(3)–Se (1), 2.6606(8) Å. *Dashed lines* indicate short P–Se nonbonding interaction at P(3)–Se(4), 3.104(2) Å. Adapted with permission from [75]. Copyright © 2007 Royal Society of Chemistry

glasses. Surprisingly, glassy $Cs_5P_5Se_{12}$ also exhibited a significant SHG response, at ~5% that of $AgGaSe_2$.

Alkali-metal salts of a novel quaternary zirconium selenophosphate phase were isolated using the molten flux technique, having a general formula of $AZrPSe_6$ (A = K, Rb, Cs) [43]. They are isostructural and crystallize in the polar space group $Pmc2_1$. Their structures feature parallel chains of the infinite $_\infty{}^1[ZrPSe_6{}^-]$ anion separated by the alkali metal ions (Fig. 11a). The chains have Zr^{4+} ions coordinated with Se atoms in a distorted bicapped trigonal prismatic geometry (Fig. 11c). All Zr^{4+} ions are connected to a $_\infty{}^1[PSe_3{}^-]$ polymeric backbone formed by the condensation of corner-sharing tetrahedral PSe_4 units (Fig. 11b). Each phosphorus atom has a terminal P–Se bond projecting out from the one-dimensional chain structure. The Zr atoms are also bridged with η^4-bonded $Se_2{}^{2-}$ groups. Finally, there is also terminal Se(3) atom on each zirconium metal center.

The polar structure and presence of highly polarizable atoms such as Se, Rb, Cs, and Zr result in their strong SHG response. The SHG response of crystalline $CsZrPSe_6$ and $RbZrPSe_6$ was ~15- and tenfold than that of the commercial NLO material $AgGaSe_2$ while the K analogue showed a weaker SHG signal intensity than $AgGaSe_2$. Phase-matching experiments performed using different size particles of $CsZrPSe_6$ indicate the material to be type I phase matchable. These materials also show wide transparency in the mid-IR region up to 18.5 μm and high solubility in hydrazine. The wide optical transparency region, high solubility along with the strong SHG response of $CsZrPSe_6$ and $RbZrPSe_6$ makes them promising for thin film infrared NLO applications.

Semiconductor $Cs_5BiP_4Se_{12}$ nanofibers with a nearly direct bandgap of 1.85 eV were synthesized under vacuum in a fused-silica tube at high temperature [76]. It adopts the polar space group $Pmc2_1$ and features discrete molecular $[Bi(P_2Se_6)_2]^{5-}$ anions (Fig. 12a), which are isolated by Cs^+ cations (Fig. 12b). The coordination environment of the Cs atoms results in a macroscopic alignment of the dipole moments along the *c*-axis, resulting in the polar nature. It melts congruently and

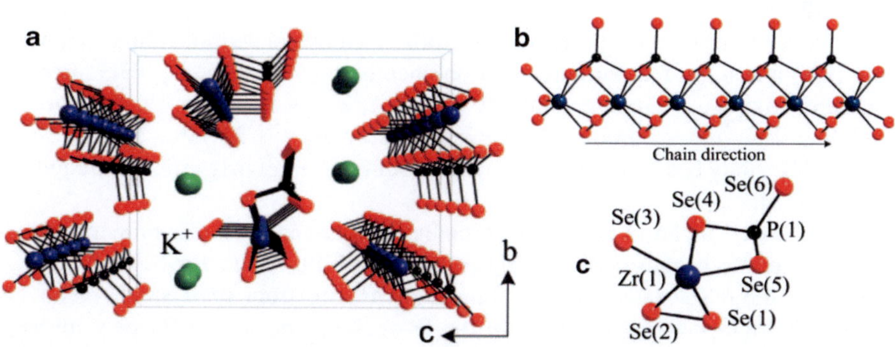

Fig. 11 (**a**) Noncentrosymmetric structure of $KZrPSe_6$: Molecular view along the [1 0 0] direction. (**b**) Polymeric $^1/_\infty[PSe_3{}^-]$ chain acting as a backbone holding the Zr^{4+} ions together. (**c**) View of $[ZrPSe_6]^-$ anion along the chain axis. Adapted with permission from [43]. Copyright © 2008 American Chemical Society

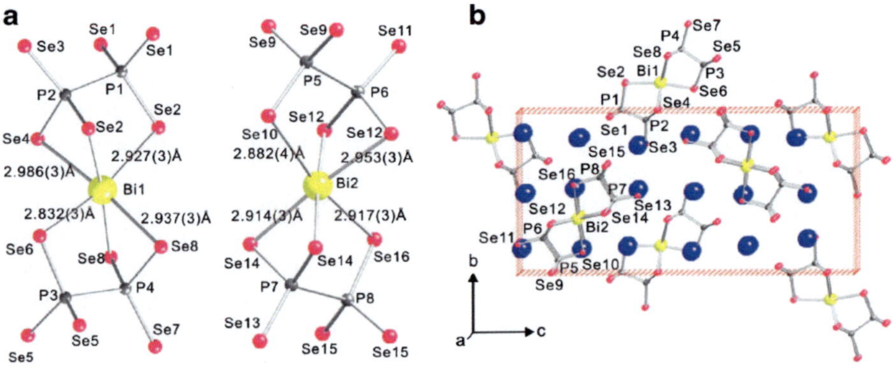

Fig. 12 (**a**) Structure of two crystallographically independent $[Bi(P_2Se_6)_2]^{5-}$ anions. Bi (*yellow*), P (*gray*), and Se (*pink*) atoms are labeled and selected bond distances (Å) shown. (**b**) Structure of $Cs_5BiP_4Se_{12}$ viewed down the *a*-axis. Bi (*yellow*), P (*gray*), and Se (*pink*) atoms are labeled. Large *blue circles* are Cs atoms. Adapted with permission from [76]. Copyright © 2009 American Chemical Society

soluble in NMF. $Cs_5BiP_4Se_{12}$ is optically transparent from the far-IR through the mid-IR to below the band edge in the visible region (18.7–0.67 μm). The SHG intensity of $Cs_5BiP_4Se_{12}$ is approximately two times larger than that of $AgGaSe_2$ at 1 μm and type I non-phase matchable.

The reaction of Cu with a molten mixture of Cs_2S/P_2S_5 produced the quaternary compound $Cs_2CuP_3S_9$ in the chiral, hexagonal space group $P6_5$ (No. 170) [77]. Its structure contains parallel helical $[CuP_3S_9]_n^{2n-}$ chains that run along the *c*-axis. The $[CuP_3S_9]_n^{2n-}$ chains are composed of an alternating arrangement of tetrahedral Cu^+ ions and cyclic $[P_3S_9]^{3-}$ units, and the cyclic $[P_3S_9]^{3-}$ unit derives from the molecular adamantane $[P_4S_{10}]$ unit with one $[PS]^{3+}$ vertex removed. Preliminary qualitative tests for NLO properties of powder samples of $Cs_2CuP_3S_9$ using a pulsed Nd:YAG laser were positive, judging from the frequency-doubled green light exiting the sample.

Lamellar ferrielectric $CuInP_2S_6$ was synthesized by conventional high-temperature (HT) and room-temperature solution-chemistry (SC) methods, SHG intensity measurements on a single-crystal HT sample reveal a transition temperature $T_c = 313 \pm 2$ K, and the frequency conversion efficiency is 26 times that of Y-cut quartz. The study has shown that crystalline $CuInP_2S_6$ is a fairly efficient frequency doubler [78].

Sulfur-based IR NLO materials normally have larger nonlinear properties than oxide materials because of the higher polarilizability of sulfur-based bonding. However, their low laser-damage thresholds arising from their small bandgaps limit their high-power application in the IR region. The search for new infrared NLO crystals with high laser-damage thresholds combined with large optical nonlinearity and wide transmission ranges, visible to IR, is therefore quite active to respond to the increasing use of infrared NLO crystals in a wide range of applications. Thioborate materials based on boron sulfide are being considered as a new class of NLO materials by combining the favorable transparency and nonlinearity of sulfur-based NLO materials (e.g., $AgGaS_2$) with the high damage thresholds of borates (e.g., LiB_3O_5).

Metal thioborates consisting of the trigonal planar $(BS_3)^{3-}$ structural unit have been investigated as new NLO materials because the planar $(BS_3)^{3-}$ unit containing conjugated π-orbital systems will produce high NLO properties and the strongly covalent bonding of boron atoms in the $(BS_3)^{3-}$ anionic group will enable high damage thresholds and a wide transparency range from the visible to mid-IR regions. A $Zn_xBa_2B_2S_{5+x}$ ($x \approx 0.2$) phase crystallized in the NCS tetragonal space group *I*-42*m* and consists of the trigonal planar $(BS_3)^{3-}$ structural units. $Zn_xBa_2B_2S_{5+x}$ ($x \approx 0.2$) yields strong SHG effects about 50 times greater than that of α-SiO_2 and has a wide transparent range from the visible to the mid-IR region (0.35–10 μm). Additionally, an absorption edge estimation supports its much higher laser-damage threshold than that of other sulfur-based NLO materials. While the $Zn_xBa_2B_2S_{5+x}$ ($x \approx 0.2$) phase is the first such NLO thioborate phase to be discovered, many other metal thioborates are possible and may yield a number of excellent candidate materials that have high nonlinear optical properties combined with wide transmission and high damage thresholds for second-order NLO application in both the visible and IR regions [44].

5.2 *Metal Pnictidohalides and Chalcohalides*

Two inorganic supramolecular compounds, $(Hg_6P_3)(In_2Cl_9)$ and $(Hg_8As_4)(Bi_3Cl_{13})$, have been prepared by mediate-temperature solid-state reaction [45]. $(Hg_6P_3)(In_2Cl_9)$ and $(Hg_8As_4)(Bi_3Cl_{13})$ crystallize in $P2_1$ and $P3_121$, respectively, both featuring chiral 3-D cationic host frameworks, $(Hg_6P_3)^{3+}$ or $(Hg_8As_4)^{4+}$, with anionic guest moieties, $(In_2Cl_9)^{3-}$ or $(Bi_3Cl_{13})^{4-}$, filling the helical tunnels (Fig. 13). The cationic network is built up from approximately linearly coordinated mercury and tetrahedrally coordinated phosphorus or arsenic atoms to form helical HgP(As) spring. Two concentric HgP(As) springs are bridged by other Hg atoms through HgP(As) bonds to form a cylinder with a 1-D helical tunnel along the *b* (or *c*) direction; then the helical cylinders share all the mercury and phosphorus (or arsenic) atoms with each other to form a 3-D honeycomb-like $(Hg_6P_3)^{3+}$ or $(Hg_8As_4)^{4+}$ cationic framework with the schäfli symbol of 5^58. $(Hg_6P_3)(In_2Cl_9)$ and $(Hg_8As_4)(Bi_3Cl_{13})$ are semiconductors with bandgaps of 3.13 and 2.43 eV, respectively; they are transparent in the whole measured region from 2.5 to 25 μm. The measured SHG signal intensities of $(Hg_6P_3)(In_2Cl_9)$ and $(Hg_8As_4)(Bi_3Cl_{13})$ are about 0.5 and 1.2 times that of $AgGaS_2$, respectively, and they are type I phase matchable. The study in the reference indicates that good NLO materials can be obtained by designing both complicated polycations and polyanions with large molecular polarizability as functional components rather than traditional single polyanions, and the spatial arrangement of the functional components can be

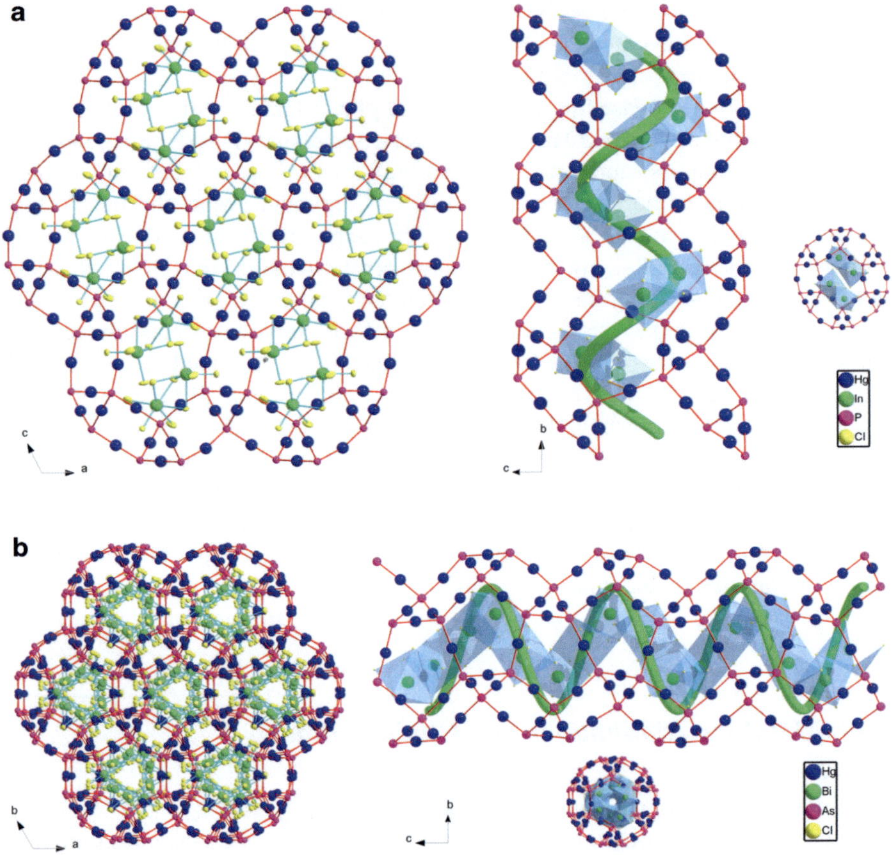

Fig. 13 View of 3-D cationic host $(Hg_6P_3)^{3+}$ framework and discrete guest $(In_2Cl_9)^{3-}$ anions of $(Hg_6P_3)(In_2Cl_9)$ along the *b* direction (**a** *left*), isolated $(In_2Cl_9)^{3-}$ anions are embedded in the helical channels of the host framework (**a** *right*). Three-dimensional cationic host $(Hg_8As_4)^{4+}$ framework and $_\infty{}^1(Bi_3Cl_{13})^{4-}$ helical chain of $(Hg_8As_4)(Bi_3Cl_{13})$ along the *c* direction (**b** *left*), $_\infty{}^1(Bi_3Cl_{13})^{4-}$ helical chain embed in the channel of the host framework (**b** *right*)

designed in a reasonable alignment to enforce macroscopic dipole ordering efficiently through crystal engineering strategy [45].

A novel chiral open-framework structural hexanary chalcogenoborate, $(K_3I)[SmB_{12}(GaS_4)_3]$, was obtained by a facile approach using boron as the reducing reagent [79]. It crystallizes in the chiral space group $P6_322$, and its structure contains B_{12}-icosahedral, distorted GaS_4 tetrahedral and distorted SmS_6 octahedral building blocks, and face-sharing IK_6 octahedral chains (Fig. 14). There are two boron, one gallium, one samarium, one potassium, and one iodine atoms in the crystallographically independent unit. The structure of $(K_3I)[SmB_{12}(GaS_4)_3]$ features B_{12} icosahedra with B1 and B2 being saturated by μ-S1 and μ_3-S2 atoms, respectively, surrounded by six GaS_4 tetrahedra and consolidated by two SmS_6 octahedra along the *c* direction to form a 3-D honeycomb-like open-framework, yielding channels which are

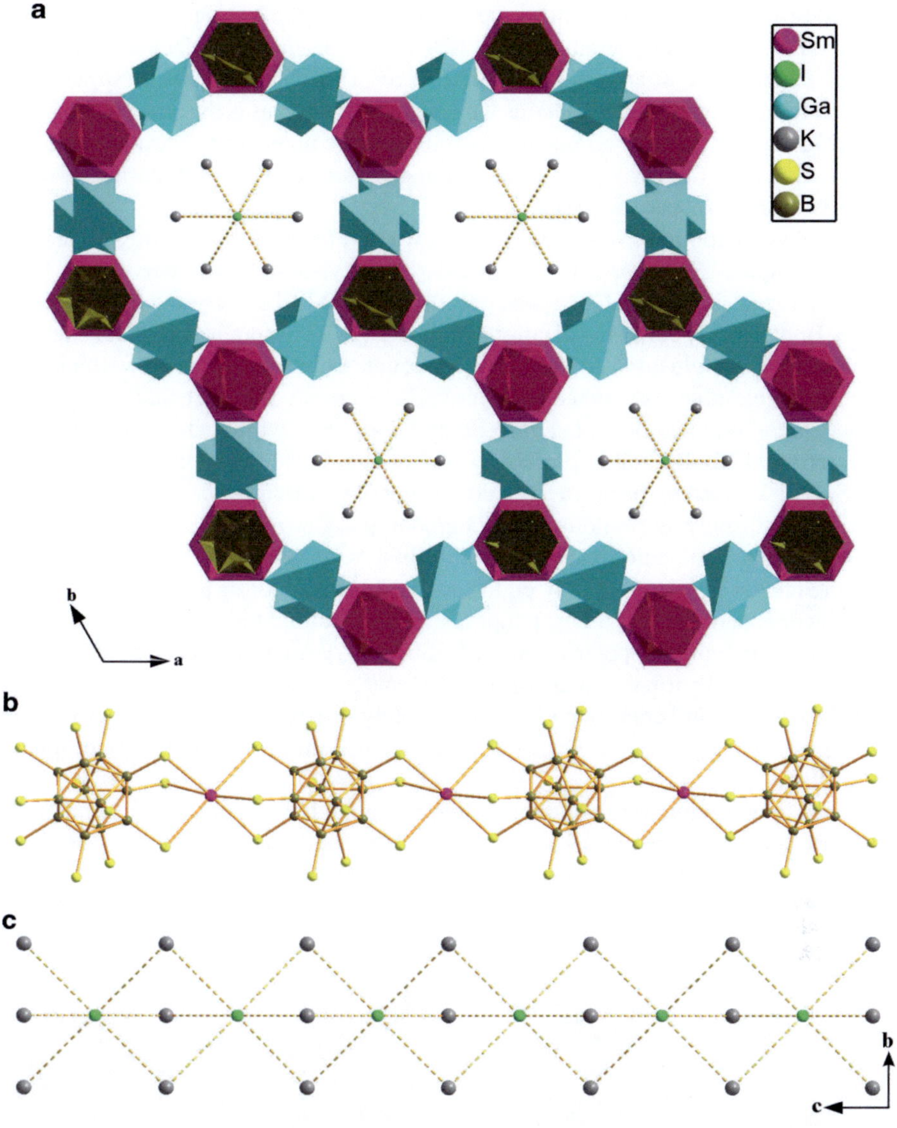

Fig. 14 (**a**) View of the structure of $(K_3I)[SmB_{12}(GaS_4)_3]$ along the *c* direction. The B_{12} icosahedra, GaS_4 tetrahedra and SmS_6 octahedra are drawn in *dark yellow*, *turquoise*, and *pink*, respectively. (**b**) The connection of B_{12} icosahedra and SmS_6 octahedra along the *c* direction. (**c**) 1-D cationic $[(K_3I)_n]^{2n+}$ chain constructed from face-sharing IK_6 octahedra along the *c* direction

occupied by face-sharing IK_6 octahedral chains along the *c* direction. SHG measurements on the unsieved powder sample irradiated by a 1,940-nm laser reveal that $(K_3I)[SmB_{12}(GaS_4)_3]$ shows a SHG efficiency of about 0.3 times that of KDP. No absorption bands are observed in the mid-IR range of 4,000–1,000 cm^{-1}

(2.5–10 μm). The diffuse-reflectance spectrum of $(K_3I)[SmB_{12}(GaS_4)_3]$ shows that it is also transparent in the region of 585–1,078 nm, and its diffuse-reflectance spectrum reveals the presence of an optical gap of 2.35 eV. $(K_3I)[SmB_{12}(GaS_4)_3]$ shows only weak SHG activity though it contains the GaS_4 tetrahedron as is the case for $AgGaS_2$, implying that the effective arrangement of SHG-active units in NCS structure leading to enhancement of macroscopic polarizabilities is important for the design of more effective NLO response materials.

The new compound $[Sb_7S_8Br_2](AlCl_4)_3$ was synthesized by reacting Sb with S in ionic liquid EMIMBr–$AlCl_3$ (EMIM = 1-ethyl-3-methylimidazolium) [80]. It crystallizes in the NCS space group $P2_12_12_1$. The crystal structure of $[Sb_7S_8Br_2](AlCl_4)_3$ consists of cationic $[Sb_7S_8Br_2]^{3+}$ clusters and $[AlCl_4]^-$ anions (Fig. 15a). Each cluster adopts a unique double-cubane structure in which two distorted cubic clusters connect by sharing one corner (the Sb1 site). The other corners are alternately occupied with Sb and S atoms. Two Sb sites (Sb2 and Sb6) have terminal Sb–Br bonds projecting out of the cluster structure. The cationic $[Sb_7S_8Br_2]^{3+}$ clusters pack in pseudohexagonally arranged columns along the *a*-axis and display a pronounced hexagonal pseudosymmetry (Fig. 15b). The voids between the cationic clusters are filled with $[AlCl_4]^-$ anions, and each cationic cluster is surrounded by eight $[AlCl_4]^-$ anions. All of the crystallographically independent $[AlCl_4]^-$ ions adopt a slightly distorted tetrahedral shape .

Optical absorption spectroscopy of $[Sb_7S_8Br_2](AlCl_4)_3$ revealed absorption edges at ~2.03 eV; it has a wide optical transparency region (3.57–7.81 μm). The SHG intensity of the compound is approximately one-third that of KDP at 700 nm and becomes comparable to that of KDP above 900 nm, and it is type I non-phase matchable.

A new quinary rare-earth sulfide–halide compound has been synthesized and its structure determined by single-crystal X-ray diffraction. $LaCa_2GeS_4Cl_3$ crystallizes in the NCS hexagonal space group $P6_3mc$ (No. 186); preliminary optical studies indicate that this material has a useful optical window extending approximately

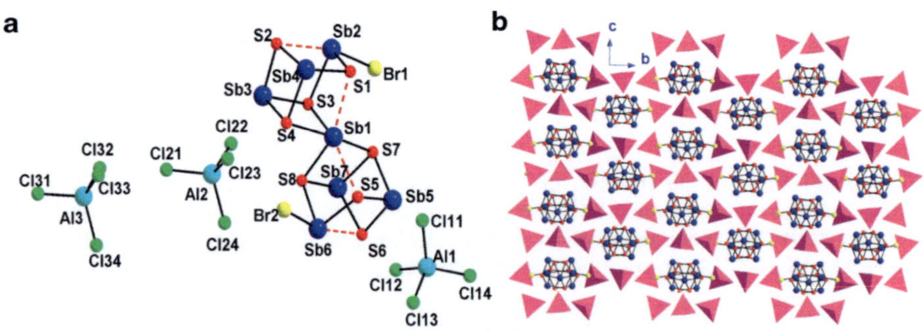

Fig. 15 (**a**) Structure of the compound $[Sb_7S_8Br_2](AlCl_4)_3$. Color code: *blue*, Sb; *red*, S; *yellow*, Br; *cyan*, Al; *green*, Cl. (**b**) Stacking view of the ions in compound $[Sb_7S_8Br_2](AlCl_4)_3$ along the *a*-axis. *Blue balls*, Sb atoms; *red balls*, S atoms; *yellow balls*, Br atoms; *purple tetrahedra*, $AlCl^{4-}$ anions. Adapted with permission from [80]. Copyright © 2009 American Chemical Society

from 0.5 to 10 μm. NLO activity of $LaCa_2GeS_4Cl_3$ is demonstrated by the generation of green light when pumped with a 1.064-μm Nd:YAG laser [81].

The structure of $LaCa_2GeS_4Cl_3$ can be simply described in terms of the metal sulfide–halide polyhedra. Individual GeS_4 tetrahedra are aligned along the threefold axis of the unit cell. All four Ge–S bond distances are nearly identical. The La/Ca sites are in sixfold, slightly distorted, trigonal prismatic coordination, with three S atoms making up one face and three Cl atoms making the other face. In this way, a typically centrosymmetric coordination becomes NCS and polar, as allowed in the 6*mm* point group of this structure. The trigonal prisms share edges on the S face and share vertices on the Cl face. The Cl atoms form a chain of face-shared, empty octahedra running up the 6_3 axis.

The optical transparency and nonlinear optical susceptibility of this material clearly indicate it is a good candidate for further studies. The demonstrated stabilities with respect to air and water, as well as reasonable thermal stability, indicate $LaCa_2GeS_4Cl_3$ could become a useful material for many optical and nonlinear optical applications.

6 Concluding Remarks

In summary, because of the indispensable application in laser devices, mid-IR NLO crystals have attracted a lot of attention from chemists and materials scientists. We have briefly reviewed the recent progress on the mid-IR NLO crystals, i.e., growing the classical mid-IR NLO crystals into large high-quality ones or into quasi-phase-matching structures that are suitable for the laser devices by various growth methods and exploring new potential mid-IR NLO crystals by introducing new design and synthesis strategies. The development of binary and ternary metal pnictides and chalcogenides is more mature than other mid-IR NLO systems in the aspect of practical applications, and some of them are commercially available, like $ZnGeP_2$, $AgGaS_2$, etc. Compared with binary and ternary phases, the compositional flexibility of quaternary materials increases and allows for the potential to tune their physical properties applied in nonlinear optics. Quaternary diamond-like metal chalcogenides, especially $AgGaGe_nS(e)_{2(n+1)}$, are a very promising mid-IR NLO material system and may become an alternative to the widely used uniaxial $AgGaS_2$ due to their higher damage threshold and greater variety of phase-matching schemes. Binary and ternary metal halides have excellent damage resistance, high nonlinear figure of merit, good mechanical properties, and high thermal conductivity coupled with transmittance over a broad spectral range that make them very promising mid-IR NLO crystals in the future. Newfound different-bond-type hybrid compounds, especially metal chalcopnictates, show very large SHG efficiencies, and some novel properties such as high solubility in organic solvent and surprising SHG response of their glass state have been found. However, lots of work must be done before they can be used in practical applications.

Although great achievements have been made on mid-IR NLO crystals, some drawbacks of recent applied crystals and difficulties in growing large crystals that hinder their wide applications still exist; advance growth techniques and equipments need to be developed in the future to grow large crystals of high quality. Exploring new potential mid-IR NLO crystals with excellent performance by introducing novel material design and synthesis strategies is another indispensable and promising direction, which can also be helpful to the establishment of the relationships between their crystal structures and mid-IR NLO performances. Two things are also worth noting here: firstly, the simpler the crystal is, the more practical it is. It may be not necessary to pursue mid-IR NLO crystals with complicated chemical compositions and structures, although they may have good mid-IR NLO properties, because it may be very difficult to grow them into large crystals, and more composition segregation and structural defects are prone to appear in them, compared with those in simple crystals. Secondly, it is no useful that only one or few mid-IR NLO properties of a mid-IR NLO crystal are excellent, while others are ordinary or bad. During the discovery of new mid-IR NLO crystals, equivalent attention should also be given to other very important attributes for SHG applications, such as high laser-damage threshold, reasonable chemical stability, and so on.

Acknowledgment We gratefully acknowledge the financial support by the NSF of China (90922035, 21003126), 973 program (2011CBA00505), Key Project from the CAS (KJCX2-EW-H03), and the NSF of Fujian Province (2010H0022).

References

1. Demtröder W (2009) Laser spectroscopy, 3rd edn. Springer, Berlin
2. Duarte FJ (ed) (2008) Tunable laser applications, 2nd edn. CRC, Boca Raton, Chapter 2, 9, and 12
3. Boyd RW (2003) Nonlinear optics. Academic, New York
4. Chen C, Liu G (1986) Annu Rev Mater Sci 16:203
5. Hussain K, Kumbhakar P (2006) Braz J Phys 36:1281
6. Yang F, Wang Z, Zhou Y, Li F, Xu J, Xu Y, Cheng X, Lu Y, Bo Y, Peng Q, Cui D, Zhang X, Wang X, Zhu Y, Xu Z (2009) Appl Phys B 96:415
7. Chen CT, Luo SY, Wang XY, Wang GL, Wen XH, Wu HX, Zhang X, Xu ZY (2009) J Opt Soc Am B 26:1519
8. Zhou Y, Wang GL, Yue YC, Li CM, Lu YF, Cui D, Hu ZG, Xu ZY (2009) Opt Lett 34:746–748
9. Chen C, Wu B, Jiang A, You G (1985) Sci Sinica (China) B 28:235
10. Chen C, Wu Y, Jiang A, Wu B, You G, Li R, Lin S (1989) J Opt Soc Am B6:616
11. Guo L, Wang GL, Zhang HB, Cui DF, Wu YC, Lu L, Zhang JY, Huang JY, Xu ZY (2007) Appl Phys B 88:197
12. Yuan X, Shen GQ, Wang XQ, Shen DZ, Wang GL, Xu ZY (2006) J Cryst Growth 293:97
13. Jing FL, Fu PZ, Wu YC, Zu YL, Wang X (2008) Opt Mater 30:1867
14. Zhang JX, Wang GL, Liu ZL, Wang LR, Zhang GC, Zhang X, Wu Y, Fu PZ, Wu YC (2010) Opt Express 18:237

15. Li F, Pan SL, Hou XL, Zhou ZX (2010) J Cryst Growth 312:2383
16. Chen C, Lin Z, Wang Z (2005) Appl Phys B 80:1
17. Eckardt RC, Masuda H, Fan YX, Byer RL (1990) IEEE J Quantum Electron 26:922
18. Zhang CL, Huang LX, Zhou WN, Zhang G, Hou HD, Ruan QF, Lei W, Qin SJ, Lu FH, Zuo YB, Shen HY, Wang GF (2006) J Cryst Growth 292:364
19. Wu YC (2002) J Synth Cryst (China) 31:288
20. Zhang SF, Zhang EP, Fu PZ, Wu YC (2009) J Cryst Growth (China) 311:2433
21. Yao SH (2009) Doc thesis, Shandong University
22. Deng Y, Zhu QH, Zeng XM, Wang FR, Huang Z, Guo Y (2007) Infrared Laser Eng (China) 36:802
23. Xue DF, Zhang SY (1998) Chem Phys Lett 291:401
24. Chen X, Liu H, Ye N (2009) J Synth Cryst (China) 38:544
25. Yang JF, Liu SD, He JL, Yang XQ, Liu FQ, Zhang BT, Xu JL, Yang HW, Huang HT (2011) Laser Phys Lett 8:28
26. Chen WC (1996) Adv Astronaut Sci 91:365
27. Chen CT, Bai L, Wang ZZ, Li RK (2006) J Cryst Growth 292:169
28. Su X, Liu T, Zhang G, Chen XG, Qin JG, Chen CT (2006) Chinese J Inorg Chem 22:1163
29. Goodman CHL (1991) Semicond Sci Technol 6:725
30. Chang CE, Wilcox WR (1974) J Cryst Growth 21:135
31. Fu TW, Wilcox WR (1980) J Cryst Growth 48:416
32. Cho AY, Arthur JR (1975) Prog Solid State Chem 10:157
33. Koh S, Kondo T, Shiraki Y, Ito R (2001) J Cryst Growth 227:183
34. Bliss DF, Lynch C, Weyburne D, O'Hearn K, Bailey JS (2006) J Cryst Growth 287:673
35. Yu XJ, Scaccabarozzi L, Lin AC, Fejer MM, Harris JS (2007) J Cryst Growth 301:163
36. Nikogosyan DN (2005) Nonlinear optical crystals: a complete survey. Springer-Science, New York
37. Dmitriev VG, Gurzadyan GG, Nikogosyan DN (1999) Handbook of nonlinear optical crystals, 3rd edn. Springer, Berlin
38. Shimamura K, Víllora EG, Muramatsu K, Ichinose N (2005) J Cryst Growth 275:128
39. Liao JH, Marking GM, Hsu KF, Matsushita Y, Ewbank MD, Borwick R, Cunningham P, Rosker MJ, Kanatzidis MG (2003) J Am Chem Soc 125:9484
40. Guo SP, Guo GC, Wang MS, Zou JP, Xu G, Wang GJ, Long XF, Huang JS (2009) Inorg Chem 48:7059
41. Zhang X, Kanatzidis MG (1994) J Am Chem Soc 116:1890–1898
42. Nguyen SL, Jang JI, Ketterson JB, Kanatzidis MG (2010) Inorg Chem 49:9098
43. Banerjee S, Malliakas CD, Jang JI, Ketterson JB, Kanatzidis MG (2008) J Am Chem Soc 130:12270
44. Kim Y, Martin SW, Ok KM, Halasyamani PS (2005) Chem Mater 17:2046
45. Jiang XM, Zhang MJ, Zeng HY, Guo GC, Huang JS (2011) J Am Chem Soc 133:3410
46. Walter RLL, Jiang XS (2004) Phys Rev B 70:045204
47. Lin X, Zhang G, Ye N (2009) Cryst Growth Des 9:1186
48. Badikov V, Badikov D, Shevyrdyaeva G, Tyazhev A, Marchev G, Panyutin V, Petrov V, Kwasniewski A (2011) Phys Status Solidi RRL 5(1):31–33
49. Komarov SA, Meleshko AN, Pleshanov AN, Solomatin VS (1980) Sov Tech Phys Lett 6:375
50. Petrov V, Rotermund F (2002) Opt Lett 27:1705
51. Ayeb Y, Ouahrani T, Khenata R, Reshak AH, Rached D, Bouhemadou A, Arrar R (2010) Comput Mater Sci 50:651
52. Poduska KM, Cario L, DiSalvo FJ, Min K, Halasyamani PS (2002) J Alloys Compd 335:105
53. Badikov V, Shevyrdyaeva G, Chizhikov V, Panyutin V, Xu GB, Petrov V, Noack F (2005) Appl Phys Lett 87:241113
54. Kim Y, Seo I, Martin SW, Baek J, Halasyamani PS, Arumugam N, Steinfink H (2008) Chem Mater 20:6048

55. Lekse JW, Moreau MA, McNerny KL, Yeon J, Halasyamani PS, Aitken JA (2009) Inorg Chem 48:7516
56. Aitken JA, Marking GA, Evain M, Iordanidis L, Kanatzidis MG (2000) J Solid State Chem 153:158
57. Poduska KM, DiSalvo FJ, Min K, Halasyamani PS (2002) J Alloys Compd 335:L5
58. Li P, Li LH, Chen L, Wu LM (2010) J Solid State Chem 183:444
59. Liu T, Qin J, Zhang G, Zhu T, Niu F, Wu Y, Chen C (2008) Appl Phys Lett 93:091102-1
60. Zhang G, Liu T, Zhu T, Qin J, Wu Y, Chen C (2008) Opt Mater 31:110
61. Ferrier A, Vela'zquez M, Doualan JL, Moncorge' R (2009) Appl Phys B-Lasers O 95:287
62. Tong YZ, Meng XY, Wang ZZ, Chen CT, Lee JMH (2005) Appl Phys 98:033504
63. Zhang J (1995) M.Sc. Thesis, Wuhan University
64. Lin ZG, Tang LC, Chou CP (2008) Opt Mater 31:28
65. Zhang G, Qin J, Liu T, Zhu T, Fu P, Wu Y, Chen T (2008) Cryst Growth Des 8:2946
66. Zhang G, Qin J, Liu T, Li Y, Wu Y, Chen C (2009) Appl Phys Lett 95:261104-1
67. Bergman JG, Chemla DS, Fourcade R, Mascherpa G (1978) J Solid State Chem 23:187
68. Velsko SP, Eimerl D (1987) J Appl Phys 62:2461
69. Feichtner JD, Roland GW (1972) Appl Opt 11:993
70. Bera TK, Song JH, Freeman AJ, Jang JI, Ketterson JB, Kanatzidis MG (2008) Angew Chem Int Ed 47:7828–7832
71. Bera TK, Jang JI, Song JH, Malliakas CD, Freeman AJ, Ketterson JB, Kanatzidis MG (2010) J Am Chem Soc 132:3484
72. Bera TK, Jang JI, Ketterson JB, Kanatzidis MG (2009) J Am Chem Soc 131:75
73. Chung I, Malliakas CD, Jang JI, Canlas CG, Weliky DP, Kanatzidis MG (2007) J Am Chem Soc 129:14996
74. Chung I, Jang JI, Malliakas CD, Ketterson JB, Kanatzidis MG (2010) J Am Chem Soc 132:384
75. Chung I, Jang JI, Gave MA, Weliky DP, Kanatzidis MG (2007) Chem Commun 4998
76. Chung I, Song JH, Jang JI, Freeman AJ, Ketterson JB, Kanatzidis MG (2009) J Am Chem Soc 131:2647
77. Hanko JA, Kanatzidis MG (2000) J Solid State Chem 151:326
78. Misuryaev TV, Murzina TV, Aktsipetrov OA, Sherstyuk NE, Cajipe VB, Bourdon X (2000) Solid State Commun 115:605
79. Guo SP, Guo GC, Wang MS, Zou JP, Zeng HY, Cai LZ, Huang JS (2009) Chem Commun 4366
80. Zhang Q, Chung I, Jang JI, Ketterson JB, Kanatzidis MG (2009) J Am Chem Soc 131:9896
81. Gitzendanner RL, DiSalvo FJ (1996) Inorg Chem 35:2623
82. Shi W, Ding YJJ, Fernelius N, Vodopyanov K (2002) Opt Lett 27:1454
83. Kuo PS, Vodopyanov KL, Fejer MM, Simanovskii DM, Yu X, Harris JS, Bliss D, Weyburne D (2006) Opt Lett 31:71
84. Petrov V, Schunemann PG, Zawilski KT, Pollak TM (2009) Opt Lett 34:2399
85. Chen CW, Hsu YK, Huang JY, Chang CS, Zhang JY, Pan CL (2006) Opt Express 14:10636
86. Rak Z, Mahanti SD, Mandal KC, Fernelius NC (2010) Phys Rev B 82:155203
87. Zhang LM, Guo J, Li DJ, Xie JJ, Andreev YM, Gorobets VA, Zuev VV, Kokh KA, Lanskii GV, Petukhov VO, Svetlichnyi VA, Shaiduko AV (2011) J Appl Spectrosc 77:850
88. Sarkisov SY, Atuchin VV, Gavrilova TA, Kruchinin VN, Bereznaya SA, Korotchenko ZV, Tolbanov OP, Chernyshev AI (2010) Russ Phys J 53:346
89. Zhang YF, Wang R, Kang ZH, Qu LL, Jiang Y, Gao JY, Andreev YM, Lanskii GV, Kokh KA, Morozov AN, Shaiduko AV, Zuev VV (2011) Opt Commun 284:1677
90. Singh NB, Suhre DR, Rosch W, Meyer R, Marable M, Fernelius NC, Hopkins FK, Zelmon DE, Narayanan R (1999) J Cryst Growth 198:588
91. Voevodin VG, Voevodina OV, Bereznaya SA, Korotchenko ZV, Morozov AN, Sarkisov SY, Fernelius NC, Goldstein JT (2004) Opt mater 26:495

92. Feng ZS, Kang ZH, Wu FG, Gao JY, Jiang Y, Zhang HZ, Andreev YM, Lanskii GV, Atuchin VV, Gavrilova TA (2008) Opt Express 16:9978
93. Das S, Ghosh C, Voevodina OG, Andreev YM, Sarkisov SY (2006) Appl Phys B 82:43
94. Hsu YK, Chen CW, Huang JY, Pan CL, Zhang JY, Chang CS (2006) Opt Exp 14:5484
95. Zhang HZ, Kang ZH, Jiang Y, Gao JY, Wu FG, Feng ZS, Andreev YM, Lanskii GV, Morozov AN, Sachkova EI, Sarkisov SY (2008) Opt Exp 16:9951
96. Myers LE, Eckardt RC, Fejer MM, Byer RL, Bosenberg WR, Pierce JW (1995) J Opt Soc Am B 12:2102
97. Lallier E, Brevignon M, Lehoux J (1998) Opt Lett 23:1511
98. Ebert CB, Eyres LA, Fejer MM, Harris JS (1999) J Cryst Growth 201/202:187
99. Eyres LA, Tourreau PJ, Pinguet TJ, Ebert CB, Harris JS, Fejer MM, Becouarn L, Gerard B, Lallier E (2001) Appl Phys Lett 79:904
100. Apollonov VV, Lebedev SP, Komandin GA, Shakir YA, Badikov VV, Andreev YM, Gribenyukov AI (1999) Laser Phys 9:1236
101. Badikov VV, Blinov PS, Kosterev AA, Letokhov VS, Malinovskii AL, Ryabov EA (1997) Quantum Electron 27:523
102. Bai L, Lin ZS, Wang ZZ, Chen CT (2008) J Appl Phys 103:083111/1
103. Isaenko L, Yelisseyev A, Lobanov S, Krinitsin P, Petrov V, Zondy JJ (2006) J Non-Cryst Solids 352:2439
104. Zondy JJ, Petrov V, Yelisseyev A, Isaenko L, Lobanov S (2008) Mid-Infrared Coherent Sources and Applications NATO Science for Peace and Security Series B: Physics and Biophysics 1:67–104
105. Zawilski KT, Schunemann PG, Setzler SD, Pollak TM (2008) J Cryst Growth 310:1891
106. Zawilski KT, Schunemann PG, Pollak TC, Zelmon DE, Fernelius NC, Kenneth Hopkins F (2010) J Cryst Growth 312:1127
107. He ZY, Zhao BJ, Zhu SF, Li JW, Zhang Y, Du WJ, Chen BJ (2011) J Cryst Growth 314:349
108. Chen BJ, Zhu SF, Zhao BJ, Zhang JJ, Huang Y, Li M, Liu J, Tan B, Wang RL, He ZY (2006) J Cryst Growth 292:490
109. Petrov V, Zondy JJ, Bidault O, Isaenko L, Vedenyapin V, Yelisseyev A, Chen W, Tyazhev A, Lobanov S, Marchev G, Kolker D (2010) J Opt Soc Am B 27:1902
110. Zhao GD, Zhu SF, Zhao BJ, Chen BJ, He ZY, Wan SQ (2009) J Cryst Growth 311:368
111. Feigelson RS (2006) J Cryst Growth 292:179
112. Ruderman W, Maffetone J, Zelman D, Poirier D (1998) Mater Res Soc Symp Proc 484:519
113. Mitra C, Lambrecht WRL (2007) Phys Rev B 76:205206
114. Rashkeev SN, Lambrecht WRL (2001) Phys Rev B 63:165212
115. Bai L, Lin ZS, Wang ZZ, Chen CT, Lee MH (2004) J Chem Phys 120:8772
116. Petrov V, Yelisseyev A, Isaenko L, Lobanov S, Titov A, Zondy JJ (2004) Appl Phys B 78:543
117. Yelisseyev AP, Drebushchak VA, Titov AS, Isaenko LI, Lobanov SI, Lyapunov KM, Gruzdev VA, Komarov SG, Petrov V, Zondy JJ (2004) J Appl Phys 96:3659
118. Zondy JJ, Vedenyapin V, Yelisseyev A, Lobanov S, Isaenko L, Petrov V (2005) Opt Lett 30:2460
119. Badikov VV, Chizhikov VI, Efimenko VV, Efimenko TD, Panyutin VL, Shevyrdyaeva GS, Scherbakov SI (2003) Opt Mater 23:575
120. Isaenko L, Vasilyeva I, Merkulov A, Yelisseyev A, Lobanov S (2005) J Cryst Growth 275:217
121. Isaenko LI, Vasilyeva IG (2008) J Cryst Growth 310:1954
122. Isaenko L, Yelisseyev A, Lobanov S, Titov A, Petrov V, Zondy JJ, Krinitsin P, Merkulov A, Vedenyapin V, Smirnova J (2003) J Cryst Res Technol 38(3–5):379–387
123. Petrov V, Isaenko L, Yelisseyev A, Krinitsin P, Vedenyapin V, Merkulov A, Zondy JJ (2006) J Non-Cryst Solids 352:2434
124. Isaenko L, Krinitsin P, Vedenyapin V, Yelisseyev A, Merkulov A, Zondy JJ, Petrov V (2005) Cryst Growth Des 5:1325

125. Isaenko L, Vasilyeva I, Yelisseyev A, Lobanov S, Malakhov V, Dovlitova L, Zondy JJ, Kavun I (2000) J Cryst Growth 218:313
126. Isaenko L, Yelisseyev A, Lobanov S, Petrov V, Rotermund F, Zondy JJ, Knippels GHM (2001) Mater Sci Semicond Process 4:665
127. Isaenko L, Yelisseyev A, Lobanov S, Petrov V, Rotermund F, Slekys G, Zondy JJ (2002) J Appl Phys 91:9475
128. Mikkelsen JC, Kildal H (1978) J Appl Phys 49:426
129. Bhar GC, Das S, Chatterjee U, Datta PK, Andreev YN (1993) Appl Phys Lett 63:1316
130. Banerjee S, Miyata K, Kato K, Saito N, Wada S (2007) Appl Phys B 87:101
131. Banerjee S, Miyata K, Kato K (2007) Opt Commun 277:202
132. Takaoka E, Kato K (1999) Opt Lett 24:902
133. Kato K, Takaoka E, Umemura N, Chonan T (2000) In: Proceedings of the conference on lasers and electro-optics Europe, CLEO Europe, paper CThE7
134. Yoshino K, Komaki H, Itani K, Chichibu SF, Akaki Y, Ikari T (2002) J Cryst Growth 236:257
135. Badikov VV, Kuz'micheva GM, Panyutin VL, Rybakov VB, Chizhikov VI, Shevyrdyaeva GS, Shcherbakov SI (2003) Inorg Mater 39:1028
136. Huang JJ, Atuchin VV, Andreev YM, Lanskii GV, Pervukhina NV (2006) J Cryst Growth 292:500
137. Andreev YM, Atuchin VV, Lanskii GV, Pervukhina NV, Popov VV, Trocenco NC (2005) Solid State Sci 7:1188
138. Eisenmann B, Jakowski M, Schaefer H (1983) Rev Chim Miner 20:329
139. Yao JY, Mei DJ, Bai L, Lin ZS, Yin WL, Fu PZ, Wu YC (2010) Inorg Chem 49:9212
140. Schunemann PG, Zawilski KT, Pollak TM (2006) J Cryst Growth 287:248
141. Knuteson DJ, Singh NB, Berghmans KA, Wagner B, Kahler D, McLaughlin S, Suhre D, Gottlieb M (2010) J Cryst Growth 312:1114
142. Reshak AH, Auluck S, Kityk IV (2008) Curr Opin Solid St M 12:14
143. Badikov VV, Don AK, Mitin KV, Seregin AM, Sinaiskii VV, Shchebetova NI (2003) Quantum Electron 33:831
144. Das S, Chatterjee U, Ghosh C, Gangopadhyay S, Andreev YM, Lanskii G, Badikov VV (2006) Opt Commun 259:868
145. Atuchin VV, Kesler VG, Ursaki VV, Tezlevan VE (2006) Solid State Commun 138:250
146. Rotermund F, Petrov V, Noack F (2000) Opt Commun 185:177
147. Petrov V, Badikov V, Panyutin V, Shevyrdyaeva G, Sheina S, Rotermund F (2004) Opt Commun 235:219
148. Badikov VV, Don AK, Mitin KV, Seryogin AM, Sinaiskiy VV, Schebetova NI (2005) Quantum Electron 35:853
149. Goryunova NA (1965) In: Anderson JC (ed) The chemistry of diamond-like semiconductors. The MIT Press, Cambridge
150. Parthé E (1964) Crystal chemistry of tetrahedral structures. Gordon and Breach Science, New York
151. Petrov V, Noack F, Badikov V, Shevyrdyaeva G, Panyutin V, Chizhikov V (2004) Appl Optics 43:4590
152. Petrov V, Badikov V, Shevyrdyaeva G, Panyutin V, Chizhikov V (2004) Opt Mater 26:217
153. Miyata D, Petrov V, Kato K (2007) Appl Optics 46:5728
154. Wang TJ, Kang ZH, Zhang HZ, Feng ZS, Jiang Y, Gao JY, Andreev YM, Lanskii GV, Shaiduko AV (2007) J Phys D: Appl Phys 40:1357
155. Badikov V, Mitin K, Noack F, Panyutin V, Petrov V, Seryogin A, Shevyrdyaeva G (2009) Opt mater 31:590
156. Das S, Ghosh C, Gangopadhyay S, Andreev YM, Badikov VV (2006) Jpn J Appl Phys 45:5795
157. Ni YB, Wu HX, Wang ZY, Mao MS, Cheng GC, Fei H (2009) J Cryst Growth 311:1404
158. Hagemann M, Weber H-J (1996) Appl Phys A 63:67
159. Huang H, Lin ZS, Bai L, He R, Chen CT (2010) Solid State Commun 150:2318

160. Bergman JG, Crane GR, Guggenheim H (1975) J Appl Phys 46:4645
161. Zhang J, Su N, Yang C, Qin J, Ye N, Wu B, Chen C (1998) SPIE 3556:1
162. Ewbank MD, Cunningham F, Borwick R, Rosker MJ, Gunter P (1997) CLEO'97 CFA 7:462
163. Lin ZG, Tang LC, Chou CP (2008) J Cryst Growth 310:3224
164. Edwards AJ (1970) J Chem Soc A 2751–2753
165. Singh NB, Suhre DR, Green K, Fernelius N, Hopkins FK (2005) J Cryst Growth 274:132
166. Ferrier A, Velázquez M, Pérez O, Grebille D, Portier X, Moncorgé R (2006) J Cryst Growth 291:375
167. Ferrier A, Velázquez M, Portier X, Doualan JL, Moncorgé R (2006) J Cryst Growth 289:357
168. Tang LC, Liu LQ, Chang YC, Yao JH, Huang JY, Chang CS (2009) Jap J Appl Phys 48:082001-1
169. Avdienko KI, Badikov DV, Badikov VV, Chizhikov VI, Panyutin VL, Shevyrdyaeva GS, Scherbakov SI, Scherbakova ES (2003) Opt Mater 23:569
170. Badikov DV, Badikov VV, Kuz'micheva GM, Panyutin VL, Rybakov VB, Chizhikov VI, Shevyrdyaeva GS, Shcherbakova ES (2004) Inorg Mater 40:314
171. Víllora EG, Shimamura K, Sumiya K, Ishibashi H (2009) Opt Express 17:12362
172. Kim SW, Chang HY, Halasyamani PS (2010) J Am Chem Soc 132:17684
173. Abrahams SC, Ravez J, Simon A, Chaminade JP (1981) J Appl Phys 52:4740–4743
174. Gu QT, Pan QW, Shi W, Sun X, Fang CS (2000) Prog Cryst Growth Charact Mater 40:89
175. Gu QT, Fang CS, Shi W, Wu XW, Pan QW (2001) J Cryst Growth 225:501
176. Gu QT, Pan QW, Wu XW, Shi W, Fang CS (2000) J Cryst Growth 212:605
177. Tang LC, Huang JY, Chang CS, Lee MH, Liu LQ (2005) J Phys Condens Matter 17:7275
178. Lin ZG, Tang LC, Chou CP (2008) Inorg Chem 47:2362
179. Lin ZG, Tang LC, Chou CP (2007) J Phys Condens Matter 19:476209
180. Ren P, Qin J, Liu T, Wu Y, Chen C (2003) Opt mater 23:331
181. Ren P, Qin J, Chen C (2003) Inorg Chem 42:8
182. Chung I, Do J, Canlas CG, Weliky DP, Kanatzidis MG (2004) Inorg Chem 43:2762

Struct Bond (2012) 145: 45–80
DOI: 10.1007/430_2011_73

Published online: 3 March 2012

Simulation and Design of Infrared Second-Order Nonlinear Optical Materials in Metal Cluster Compounds

Kechen Wu

Abstract In this minireview, we overview the recent advances and perspectives in the developments of the infrared second-order nonlinear optical materials. The traditional semiconductors are discussed first including the problems encountered such as the facility of large second-order nonlinearity but difficulty in practical materials for laser applications. We then focus our special interest on the area of the transition-metal polynuclear cluster compounds which is a great promising area for developing new-generation infrared second-order nonlinear optical materials and molecule-scaled photoelectronic devices. We present in detail the computational studies on the microscopic mechanism of second-order nonlinear optical response and the structure–property relationship insight of these metal cluster compounds.

Keywords Charge transfers · Density functional calculations · IR absorption · Nonlinear optics · Transition-metal compounds

Contents

K. Wu (✉)
State Key Laboratory of Structural Chemistry, Fujian Institute of Research on the Structure of Matter, Chinese Academy of Sciences, Fuzhou 350002, People's Republic of China
e-mail: wkc@fjirsm.ac.cn

1 Introduction of IR NLO Materials

The infrared (IR) optical crystalline material is optically transparent in the IR region (0.7–400 μm), i.e., it is not absorbed or less absorbed by the light in this specified spectroscopic region. In fact, most of the materials applied in the IR region cannot be transparent through the whole IR region. The materials that have one or more applicable transparent windows are named as IR optical materials specified in their transparency windows. The linear optical (LO) modulations of the IR optical materials such as propagation, reflection, refraction, diffraction, etc., have been widely applied in the areas of meteorological monitoring, remote sensing, laser distance measuring, missile guidance, and laser communication. The IR optical material with the large nonlinear optical (NLO) response (so-called IR NLO material) could perform NLO modulations to light, especially laser light, in the IR region. It can nonlinearly modulate the laser frequency like frequency doubling, frequency converting, and frequency oscillating [1]. At present, the laser-pumped radiations in the IR region are less than those in the UV-vis region (1.064 μm radiation pumped by Nd-YAG laser in the near-IR region and 10.6 μm pumped by CO_2 gas laser in the mid-IR region are two typical ones). The applicable IR NLO materials can effectively widen the IR laser radiations so that they can greatly improve the applications of laser devices in the long-wavelength space. The most demanded example is long-distance laser communication, in which the signal-encoded laser light with longer wavelength is necessary due to diffraction. The coherent laser beams transparent within 5–15 μm region are preferable. Consequently, it is demanded to develop the high-efficient, easy-used, and low-cost mid-to-far IR laser devices, in which the IR NLO materials play the key role. The special advantage of the crystalline solid IR NLO materials lies in the production of all-solid state IR laser devices, which are portable and stable in practical applications.

In 1971, Chemla demonstrated the second harmonic generation (SHG) effect of silver thiogallate ($AgGaS_2$) crystal in the IR region [2]. $AgGaS_2$ can be optically transparent from 0.47 μm to 13 μm, and it belongs to $\bar{4}2$ m point group and it is a negative uniaxial crystal. The second-order nonlinear coefficient measured by Boyd showed that d_{36} at 10.6 μm is 11.1 pm/V [3], which is about 30 times larger than the d_{36} coefficient at 1.064 μm of the typical UV-vis NLO crystal, potassium dihydrogen phosphate (KDP). Up to now, $AgGaS_2$ crystal is the most common and representative IR NLO crystal and has been widely used in many areas [4, 5]. $AgGaS_2$ crystal is a typical I–III–VI chalcopyrite semiconductor with an energy gap of 2.7 eV. The laser-induced surface damage threshold (I_{thr}) at 10.6 μm is 0.1–0.2 × 10^{-12} W/m^2 with 150 ns pulse [6]. As a comparison, the laser damage threshold of KDP crystal at 1.064 μm (stronger laser beam) is 3–6 × 10^{-12} W/m^2 with 20 ns pulse [5]. The lower laser damage threshold limits the IR applications of $AgGaS_2$ crystal because it is easier to be destroyed by the high-power laser beam (basically due to the instant multiple-photon absorptions and/or Raman resonance absorptions). And thus, the laser devices produced by it are less durable.

Zinc germanium phosphite ($ZnGeP_2$) is another ternary chalcopyrite semiconductor from II–IV–V group with the same point group and similar structure as $AgGaS_2$ crystal [7]. It has a much larger NLO coefficient of $d_{36} = 68.9$ pm/V at 10.6 μm [8], which is about seven times larger than that of $AgGaS_2$ crystal. It is transparent in the region of 0.65–13 μm, where it has sufficient birefringence for phase-matching. It has good mechanical property as well with Mohs hardness being 5.5. But its optical damage threshold is also very low with I_{thr} being 0.78×10^{-12} W/m^2 at 9.6 μm with 120 ns continuum laser pulse [9]. Furthermore, a stable growth of single crystal with high structural perfectibility and uniformity is very difficult [10], which seriously limits the IR applications in laser devices.

Therefore, the two important factors that restrict the applications of IR second-order NLO semiconductors are (a) the low optical damage threshold and (b) the difficulty in perfect single crystal growth. Although the studies of the UV-vis NLO crystals also encounter the two problems, the small energy gap and the necessary optical transparency in the IR region are specified to the IR NLO crystals.

The great efforts have been made in the last decades to overcome the above-mentioned limitations of the IR second-order NLO semiconductors. The element substitution is one of the methods of choice. Silver gallium selenide ($AgGaSe_2$) was demonstrated as the IR second-order NLO crystal by Chemla et al. [11]. It is another ternary semiconductor with chalcopyrite structure similar to $AgGaS_2$, but the sulfur element has been substituted by selenide. Comparing to $AgGaS_2$ crystal, it widens the optical transparency window up to 19 μm, enhances d_{36} (10.6 μm) coefficient to 33 pm/V, which is about three times larger than that of $AgGaS_2$ crystal. The substitution effect is obviously positive. However, the laser-induced surface damage threshold of $AgGaSe_2$ crystal is still small since the measured I_{thr} is 0.1–0.2 $\times 10^{-12}$ W/m^2 at 10.6 μm with 150 ns continuum pulse [12]. Many examples can be found in the literatures.

In the recent years, besides the development of the novel growth techniques for perfect single crystal, the research focuses mostly on the enhancement of the laser damage threshold of the IR second-order NLO materials for practical applications. Although the relationship between the structure and optical damage threshold has not yet been clearly understood, some reports revealed that the enlarged energy gap (ΔE) could enhance I_{thr} because the larger ΔE would make the multiple photon absorption more difficult to be happened [13]. The larger ΔE could be obtained by introducing light elements into the compounds. One example is the substitution of the heavier transition-metal ions (like silver) by the alkali or alkaline metal ions in chalcopyrite semiconductors. Another choice is the substitution of sulfur or selenide by halogen group elements which possess the lone-pair electrons benefiting to the large second-order NLO effect. For instance, $NaSb_3F_{10}$ crystal reported by Qin and Chen [14] was discovered in terms of the above-mentioned two considerations. The I_{thr} of this crystal has been reported to be much larger than those of chalcopyrite semiconductors. The problem is that the IR transparent edge of this crystal has been reduced and the SHG effect has been decreased. The essential reason relates to the transparency–nonlinearity trade-off relationship. The enlargement of ΔE would intrinsically reduce the d coefficient because they are inversely propositional to each other.

By the element substitution and structural modification, people have effectively tuned the SHG coefficients, transparency windows, and I_{thr} values. This structural tuning would be an important means in the further development of novel IR NLO materials in semiconductor family. Nevertheless, the transparency–nonlinearity trade-off incarnates in nonlinearity and optical damage threshold as well. In other words, the difficult choice between I_{thr} and d coefficient makes the space very limited of the discovery of the practical IR second-order NLO materials in semiconductors. It is obvious that new research area needs to be explored for developing the new-generation IR second-order NLO materials.

The IR optical materials are usually characterized by containing heavy elements. In the past decade, the second-order NLO materials containing transition-metal elements have attracted great interest. Di Bella in 2001 overviewed the transition-metal complexes (organometallic and coordination) as second-order NLO materials and pointed out that the transition-metal complexes offer a very large variety of structure and can satisfy different aspects of second-order NLO materials [15].

The studies on inorganic metal cluster compounds as the IR second-order NLO materials have been carried out in our research group for more than ten years, particularly by using the first principle computational studies on the structure–property relationship and material designing and simulations. The studies revealed that the polynuclear metal cluster compounds containing direct metal–metal bonding is a promising area in searching novel mid-to-far IR second-order NLO materials. Besides the various structures, high stability and diverse electronic property, the prior advantage of the metal cluster compounds is the metal-to-metal charge transfer contribution to the second-order NLO response. This advantage makes it possible to separate the optical absorption (usually caused by ligands in near-to-mid IR region) from the NLO response (could significantly caused by metal–metal interactions), which benefits to the solution of transparency–nonlinearity trade-off dilemma in semiconductor family. In this review article, we will present the recent advances in the first-principle computational studies on the second-order NLO properties of polynuclear cluster compounds and the elucidation of the structure–property relationship for the purpose of discovering new practical second-order IR NLO materials.

2 Metal Cluster Compounds for Second-Order Nonlinear Optics

The high value of molecular quadratic hyperpolarizability (β) is the prerequisite of a material to have the strong second-order NLO effect. The great efforts have been made on developing different NLO chromophores with high quadratic hyperpolarizability [16–24]. By introducing the heavy transition metal into a pure organic donor–π–acceptor conjugated molecular architecture, the electron push–pull mechanism could be reinforced which benefits to the high molecular quadratic hyperpolarizability. Since the pioneering works of Frasier [25] and Green [26], the discovery has aroused the great interest in developing organometallic NLO chromophores for two decades (for a contribution before 1991, see, e.g., [27] and the references therein; for a contribution before 2001, see, e.g., [15] and the references therein) [28–34]. Some organometallic complexes have been found to have extremely large quadratic hyperpolarizabilities, good photochemical stability, comprehensive charge transfer ability, and tailoring flexibility of coordination ligand. The extensive studies have provided profound understanding of the mechanism for the second-order response of the organometallic [35–39] and inorganic mononuclear complexes [40–44]. The dominant second-order NLO mechanism of the organometallic chromophores lies in the intense, low-energy charge transfers between the metal centers and coordination ligands (MLCT/LMCT) [45–51]. In some particular condition, the metal-inductive ligand-to-ligand charge transfers (LLCT) or intraligand charge transfers (ILCT) can play the key role in second-order activities [52, 53] as well. These revealed that second-order NLO mechanisms all critically depend on the coordination ligands (L) which leads to the ineluctable exploitation on the extended conjugated ligands in developing organometallic NLO chromophores. For example, Le Bozec and his coworkers reported an octupolar zinc complex exhibiting very high quadratic hyperpolarizability ($\beta_{1.91} = 870 \times 10^{-30}$ esu and $\beta_0 = 657 \times 10^{-30}$ esu) [54]. The coordination ligand of this compound is the extended long ligand, 4,4′-oligophenylenevinylene-functionalized 2,2′-bipyridine which is responsible to the large β value originated from the MLCT process. Both experimental and theoretical studies came to the similar conclusion that the nature of coordination of conjugated ligands crucially influences the second-order response [23, 29–31, 55]. However, some reports further indicated the unfavorable red shifts of the low-lying charge transfer excitations into the visible part or even into the near-to-mid IR part of the spectra and the large dipole moments unfavorable to the noncentrosymmetric crystallization due basically to the extended π-conjugated coordination ligands. These drawbacks seriously restrict the practical nonresonant applications of large numbers of the organometallic NLO compounds in the IR region. The present challenge is to explore the metal-based NLO chromophores with an improved transparency–nonlinearity trade-off [56–58], which requires to simultaneously control the primary NLO property (the quadratic hyperpolarizability) and the secondary property (optical transparency) through the structural tuning. The potential new mechanism for the second-order response that is *independent* on the size and order of the coordination ligands is largely expected.

Some studies have implied the contribution to the optical nonlinearities of the *direct* metal–metal interaction of inorganic polynuclear metal clusters compounds [59–62]. In contrast to the organometallic chromophores, the charge transfer (CT) processes of polynuclear metal cluster compounds involved in the NLO response are notably diverse and complex. The metal–metal interaction CT process (MMCT) activated by the incident laser light related to the d_1–d_2 transitions makes significant contributions to the second-order NLO activities, which will be discussed in detail in the following sections. The understanding of this distinct contribution will cast a new light on the rationalization of the role of the *direct* metal–metal interactions in global NLO activity and will benefit to the exploration of the novel IR second-order NLO materials [63]. This *ligand-independent* mechanism for the second-order response can bring in an unprecedented way in tuning the quadratic hyperpolarizabilities of metal clusters in addition to the modifications of the ligating groups. It is thus critically important to understand the mechanisms and the role that transition metals play in the enhancement of the quadratic hyperpolarizability in the development of the effective NLO-active chromophores which are essential in the design of novel IR second-order NLO materials.

2.1 Computational Method for Metal Cluster Compounds

Many theoretical efforts have been made on the second-order NLO mechanism of the organometallic complexes in the past two decades both at the semiempirical level and in recent years within the first-principle theoretical frameworks [31, 35, 36, 46, 64–67].

The calculations of quadratic hyperpolarizability require the high-level computational techniques such as large basis sets and electronic correlation correction in order to compare with the experimental results. In such cases, the fast-developed density functional theory (DFT) becomes the method of choice for the transition-metal-containing complexes owing to its ability to deal with the sizable molecules and to take into account the complex many-body effects at an economic computational cost. However, some studies carried out by DFT studies reported the failures in computing quadratic hyperpolarizabilities of the long molecules and extend organometallic complexes because the DFT results tend to underestimate the long-range electronic excitations [68–73]. For example, Bruschi et al. compared ab initio and DFT calculations of the organometallic carbonyl complexes of M $(CO)_5$ L (M = W, Cr, L = Py, PyCHO, Pyz, $PyzBF_3$, BPE, $BPEBF_3$), which showed the serious overestimation of the DFT calculations on the quadratic hyperpolarizability [74]. The chief point lies in the lack of the *exact* general exchange–correlation (XC) functional to describe the various chemical phenomena. The great efforts have been made to rectify the defect within the DFT framework in order to obtain the reliable quadratic hyperpolarizability of the long molecules, oligomers, and extended polymeric systems [75–79] The Becke's parameter-fitting hybrid methods which take into account the HF "exact" exchange based on the adiabatic connection method (ACM) is one of the pathways to improve the DFT calculations of the quadratic hyperpolarizabilities. We described in the following subsection a new computational approach based on hybrid DFT (*c*PW1PW91) benchmark which is reliable to the quadratic hyperpolarizability of transition-metal-containing systems.

2.1.1 One-Parameter Hybrid Functional: *c*PW1PW91

According to Kohn–Sham density functional theory, the molecular electronic energy could be divided into several terms:

$$E = E_{\mathrm{T}} + E_{\mathrm{V}} + E_{\mathrm{J}} + E_{\mathrm{XC}} \tag{1}$$

where E_{T}, E_{V}, and E_{J} are the kinetic energy, potential energy, and electron–electron repulsion energy terms, respectively. The nonclassical energy term E_{XC} accounts for the exchange energy arising from the antisymmetry of the wave function and the correlation of the individual electrons. E_{XC} could be divided into two parts namely the exchange and correlation parts:

$$E_{\mathrm{XC}} = E_{\mathrm{X}} + E_{\mathrm{C}} \tag{2}$$

The hybrid E_{XC} functional with the fractional HF exchange and DF exchange along with DF correlation formulated by Becke has the general form as follows:

$$E_{\mathrm{XC}}^{\mathrm{hybrid}} = P_1 E_{\mathrm{X}}^{\mathrm{HF}} + P_2 E_{\mathrm{X}}^{\mathrm{DFT}} + P_3 E_{\mathrm{C}}^{\mathrm{DFT}} \tag{3}$$

To the case of Beck-style three-parameter functional (B3LYP) [80, 81] formula, 3 is written as follows:

$$\begin{aligned} E_{\mathrm{XC}}^{\mathrm{B3LYP}} = {} & E_{\mathrm{X}}^{\mathrm{LDA}} + a_0\left(E_{\mathrm{X}}^{\mathrm{HF}} - E_{\mathrm{X}}^{\mathrm{LDA}}\right) + a_{\mathrm{X}}\Delta E_{\mathrm{X}}^{\mathrm{Beck88}} + E_{\mathrm{C}}^{\mathrm{VWN}} \\ & + a_{\mathrm{C}}\left(E_{\mathrm{C}}^{\mathrm{LYP}} - E_{\mathrm{C}}^{\mathrm{VWN}}\right) \end{aligned} \tag{4}$$

where the parameter a_0 represents the ratio between the HF exchange and LDA local exchange, while a_{x} scales Beck88 gradient correction to LDA exchange. Similarly, the parameter a_{c} weights the LYP nonlocal correlation correction to the local VWN correlation functional. The three fractional parameters $\{a_0, a_{\mathrm{x}}, a_{\mathrm{c}}\}$ $(0 \leq \{a_0, a_{\mathrm{x}}, a_{\mathrm{c}}\} \leq 1)$ have been determined by fitting the G1 molecule set: $a_0 = 0.20$, $a_{\mathrm{x}} = 0.72$, and $a_{\mathrm{c}} = 0.81$. Unfortunately, B3LYP method gave bad results to the quadratic hyperpolarizabilities of long organic molecules reported in the literatures [70] and the metal complexes as well, due basically to the parameter optimization being performed in ground-state situation.

The new hybrid E_{XC} functional, on the basis of the assessments of various XC functional models, adopted the *m*PW model (Barone's modified Perdew–Wang 1991 exchange functional [82]) as the nonlocal exchange and the PW91 (Perdew–Wang gradient-corrected correlation functional [83]) for nonlocal correlation functional. Both the local exchange and correlation functionals used LDA model, Slater style for exchange, and VWN for correlation. Formula 4 comes to the following form:

$$E_{\mathrm{XC}}^{\mathrm{hybrid}} = E_{\mathrm{X}}^{\mathrm{Slater}} + a_0\left(E_{\mathrm{X}}^{\mathrm{HF}} - E_{\mathrm{X}}^{\mathrm{Slater}}\right) + a_{\mathrm{X}}\Delta E_{\mathrm{X}}^{\mathrm{mPW}} + E_{\mathrm{c}}^{\mathrm{VWN}} + a_{\mathrm{c}}\left(E_{\mathrm{c}}^{\mathrm{PW91}} - E_{\mathrm{c}}^{\mathrm{VWN}}\right) \quad (5)$$

Champagne et al. reported that for the long organic molecules, the correlation part in an XC functional has negligible effect on β, while the exchange part is mainly responsible [68]. Our results also showed that the a_0 is much more sensitive to β than the other two ones [73]. Consequently, it is reasonable to simplify the three-parameter fitting to the one-parameter fitting, i.e., $\{a_0, a_c\}$ are set to constant, $a_x = 1, a_c = 1 - a_0$. The minimum value of the mean absolute deviation (MAD, *S*) gives rise to the optimized a_0 parameter. Formula 5 came to the following:

$$E_{\mathrm{XC}}^{\mathrm{hybrid}} = E_{\mathrm{X}}^{\mathrm{Slater}} + a_0\left(E_{\mathrm{X}}^{\mathrm{HF}} - E_{\mathrm{X}}^{\mathrm{Slater}}\right) + (1 - a_0)\Delta E_{\mathrm{X}}^{\mathrm{mPW}} + E_{\mathrm{C}}^{\mathrm{PW91}} \quad (6)$$

The one-parameter a_0 has been determined by the optimization of the quadratic hyperpolarizability of a sample set consisting of 27 transition-metal-containing molecules. Due to the lack of the experimental results of the quadratic hyperpolarizability of the sample molecules, the CCSD/6-311+G(d)/SDD results were used as the gauge in the parameter a_0 fitting process.

Figure 1 showed the curve of MAD values (*S*) of the 27-molecule sample set with respect to a_0 parameter. The minimum *S* value was found at $a_0 = 0.40$. As the result, formula 6 became

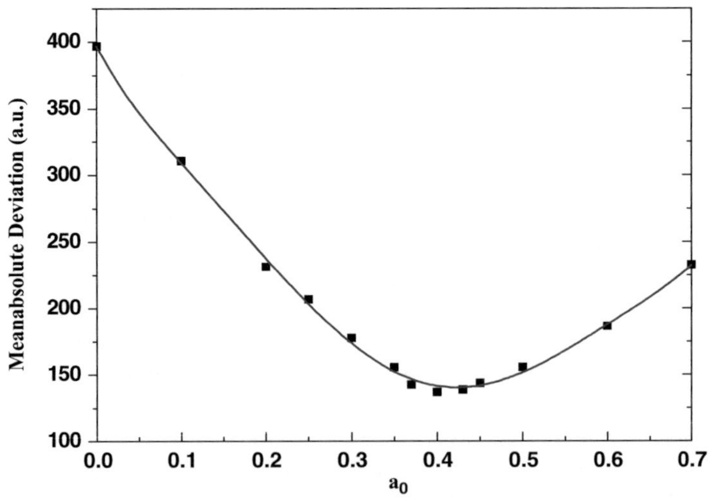

Fig. 1 Mean absolute deviation (MAD) of β values of the 27-moleucle sample set by fitting the parameter a_0. (This material is reproduced with permission of John Wiley & Sons, Inc. J Computational Chemistry 2009, 30, p.2061. ©[2009 Wiley Periodicals, Inc.])

Table 1 Calculated static quadratic hyperpolarizabilities (10^{-30} esu) of typical organometallic NLO chromophores using various computing models. (This material is reproduced with permission of John Wiley & Sons, Inc. J Computational Chemistry 2009, 30, p.2058. Copyright [2009 Wiley Periodicals, Inc.])

XC	$W(CO)_5PyCHO$	$Cr(CO)_3Bz$	$FeCp_2COMe$
SVWN	−5.52 (+47)	−25.43 (+179)	−0.96 (+38)
BLYP	−5.99 (+60)	−24.91 (+174)	−1.18 (+69)
BB95	−5.74 (+53)	−24.22 (+166)	−1.00 (+43)
BPW91	−5.67 (+51)	−24.04 (+164)	−0.95 (+36)
PW91PW91	−5.82 (+55)	−24.28 (+167)	−1.19 (+70)
mPWPW91	−5.75 (+53)	−24.16 (+165)	−1.04 (+49)
PBEPBE	−5.79 (+54)	−24.22 (+166)	−1.06 (+52)
mPW1PW91	−3.87 (+13)	−13.98 (+54)	−0.87 (+24)
B3PW91	−4.27 (+14)	−16.03 (+76)	−0.83 (+18)
B1LYP	−4.18 (+11)	−14.53 (+60)	−0.95 (+36)
B3LYP	−4.36 (+16)	−16.60 (+82)	−0.96 (+37)
M05	−4.11 (+10)	−14.74 (+62)	−1.00 (+43)
PBE1PBE	−4.28 (+14)	−14.08 (+55)	−0.84 (+20)
B97-2	−4.20 (+12)	−15.48 (+70)	−0.85 (+22)
BHandHLYP	−2.63 (−30)	−7.1 (−26)	−1.03 (+47)
HF	−1.28 (−66)	−2.19 (−75)	−0.32 (−54)
MP2	−3.13 (+17)	−6.43 (−29)	−1.72 (+146)
*c*PW1PW91	−3.92 (+5)	−9.28 (+2)	−0.80 (+14)
Expt.	−3.75	−9.10	−0.70

$$E_{\mathrm{XC}}^{\mathrm{hybrid}} = E_{\mathrm{X}}^{\mathrm{Slater}} + 0.40\left(E_{\mathrm{X}}^{\mathrm{HF}} - E_{\mathrm{X}}^{\mathrm{Slater}}\right) + 0.60\Delta E_{\mathrm{X}}^{\mathrm{mPW}} + E_{\mathrm{C}}^{\mathrm{PW91}} \tag{7}$$

Table 1 showed that this new one-parameter hybrid functional (named as *c*PW1PW91) could significantly improve the DFT performance in the computation of the quadratic hyperpolarizability of three typical organometallic complexes, $W(CO)_5PyCHO$, $Cr(CO)_3Bz$, and $FeCp_2COMe$. The improved accuracy indicates that the exact exchange functional is important in the description of excitation and NLO properties. This would illuminate the further development of the analytical XC functional in describing the electronic excitations and the NLO properties of the transition-metal-containing systems.

2.1.2 Orbital-Decomposition Analysis for Second-Order NLO Mechanism

The reliable theoretical calculations on the quadratic hyperpolarizabilities provide the important information in the analysis and understanding of the microscopic origin, NLO response mechanism, and structure–property relationship, which is critical to design and discover the novel second-order NLO materials.

The widely used two-level model comes from the "sum-over-state" method based on the perturbation theory, where the sum is simplified to only the ground state and a single excited state. It takes into account a particular excitation in analyzing the contribution to static β of the electronic transition induced by the charge transfer.

$$\beta_{\mathrm{CT}} \propto \frac{\Delta\mu_i M_i^2}{\Delta E_i^2} \tag{8}$$

where subscript i denotes the specified ith excited state. The $\Delta\mu_i$ is the difference of transition dipole moment between the ith excited state and the ground state. M_i is the transition dipole moment and ΔE_i is the transition energy from the ground state to the ith excited state. The two-level model still seems to be the useful and practical design rule for the NLO chromophores. However, the studies on the transition-metal-containing systems have been revealed that the two-level model is *invalid* any more to the extended and complex systems.

The orbital decomposition scheme firstly proposed by Hieringer and Baerends [84] is a very useful tool to relate the electronic structures and the relevant orbital-pair transitions to the quadratic hyperpolarizability. We have improved this decomposition to understand the second-order NLO nature of organometallic chromophores and coordination complexes. It can collect many important contributions which are missed by using the simple two-level model. The β value induced by the CT involving the relevant orbital-pair transitions was described as follows:

$$\beta_{(a,b)}^{\mathrm{CT}} \propto \sum_{i}^{N} \frac{\Delta u_i c_{(a,b)} M_{(a,b)}^2}{\Delta E_i^2} \tag{9}$$

where (a, b) is an *occupied* to *virtual* orbital-pair transition involved in the ith excitation; $M_{(a,b)}$ is the corresponding transition dipole moment; $C_{(a,b)}$ is the weight factor of (a, b) transition. The summation is over N states that possess the (a, b) orbital pair. The $\Delta\mu_i$ could be obtained by using the finite-field method [66]. By using this scheme, the relative contribution to β_{CT} of a specified orbital-pair transition with respect to that of another one can be clearly shown.

Another advantage of the decomposition scheme lies in the economic computational burden. It could be obtained as a by-product of the quadratic hyperpolarizability calculations without any additional computations. So it is particularly useful to the complex sizable systems like the polynuclear NLO chromophores.

2.2 Second-Order NLO Properties of Dinuclear Rhenium Clusters

We present in this subsection the study of the metal–metal interaction effect on the quadratic hyperpolarizabilities of two dinuclear rhenium clusters, $Re_2(allyl)_4$ and $Re_2(\mu\text{-}S)_2O_2(CH_2CMe_3)_4$. The electronic structures, electronic excitation, and quadratic hyperpolarizabilities have been computed and analyzed with the use of the high-level DFT/TDDFT methods. The geometries and the first intense excitations agree with the relevant reported measurements. An unprecedented second-order NLO response mechanism was found and discussed in these dirhenium compounds featuring the contribution of the direct metal–metal interaction transition process. This contribution positively enhances the quadratic hyperpolarizability and relates to the intensity of the metal–metal interactions of the complexes.

2.2.1 Structures and Computational Details

The initial geometric structures of the two dirhenium models taken from the X-ray diffraction data [85, 86] (Fig. 2) were fully optimized with C_2 symmetric restriction to the local energy minima which have been confirmed by no imaginary harmonic vibration frequency. The twofold axes of both models were arranged to be along Cartesian z axes which are also identical to their dipole moment directions.

The geometric optimizations and ground-state self-consistent-field (SCF) calculations were proceeded at the triple-ζ split-valence Slater orbital-type basis set with two augumented polarization functions (TZ2P) and the "small" frozen core level: (Re:4*d*; O:2*s*; S: 2*s*; N:2*s*). The Becke–Perdew (BP) pure general gradient approximation (GGA) XC functional [87, 88] was used with the local density approximation (LDA) part being VWN type [89, 90] including the Stoll correction [91]. The scaled-ZORA Hamiltonian [92–94] was used to take account of the relativistic effect in the calculations. All the calculations were performed with the Amsterdam density functional program (ADF) [95, 96].

The orientationally averaged static values of the polarizability (α) and quadratic hyperpolarizability (β) are defined as follows:

$$\bar{\alpha} = \frac{1}{3}\left(\alpha_{xx} + \alpha_{yy} + \alpha_{zz}\right) \tag{10}$$

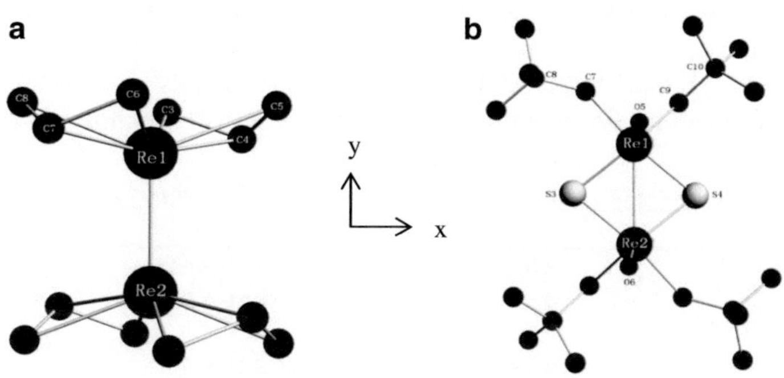

Fig. 2 Molecular structures of the models **1** and **2** (**a**) $Re_2(allyl)_4$; (**b**) $Re_2(\mu\text{-}S)_2O_2(CH_2CMe_3)_4$. H atoms are omitted for clarity

$$\overline{\beta} = \sqrt{\sum_i \beta_i^2} \quad (11)$$

$$\beta_i = \frac{3}{5} \sum_{j=x,y,z} \beta_{ijj}, \quad i = x, y, z \quad (12)$$

Since the lack of the measured data available for the β of the two dirhenium complexes, the present computed values were largely qualitatively valuable. Based on this consideration, the solvent effects and intermolecular interactions, as well as frequency-dependent effects, which have been proved to be quantitatively important, are not included in the present qualitative evaluations of the quadratic hyperpolarizability.

2.2.2 Results and Discussions

Electronic Structures and Re–Re Bonding

The mean Re–Re bond distance in $Re_2(allyl)_4$ molecule (**1**) is 2.225 Å which was considered effectively triple by Cotton et al. [85]. The Re centers in the $Re_2(\mu\text{-}S)_2O_2(CH_2CMe_3)_4$ molecule (**2**) have square-pyramidal geometries with the oxo ligand in the apical position and the basal plane defined by two carbons and bridging sulfide group. The Re–Re distance of 2.759 Å is slightly longer than the sum of two Re atomic radii (2.741 Å), which was considered as single bond by Hoffman et al. [86]. The optimized geometric parameters are in reasonable agreement with the experimental data.

Mayer bond-order method (Table 2) [97] was adopted to analyze the bond properties. The bond order of Re–Re of model **1** is 2.20. On the contrary, the calculated Re–Re bond order is 0.65 indicating the much weak interaction between two rhenium atoms in model **2**. The results show the strong Re–Re bonding in model **1** and the weak Re–Re interaction in model **2**, which agree with the experiments.

The frontier molecular orbitals of model **1** are mostly composed of the 5*d* orbitals of dirhenium. For example, the HOMO (abs. as H) mainly locates on two

Table 2 Selected bond orders of models **1** and **2** obtained by Mayer method

1		**2**	
Re1–Re2	2.00	Re1–Re2	0.65
Re1–C3	0.33	Re1–S3	1.01
Re1–C4	0.58	Re1–S4	0.93
Re1–C5	0.33	Re1–O5	1.79
Re1–C6	0.33	Re1–C7	0.74
Re1–C7	0.58	Re1–C9	0.77
Re1–C8	0.33		

Re atoms with typical δ metal–metal antibonding orbital character (noted as $d_{mm\delta*}$) in agreement with the semiempirical result of Cotton et al. The HOMO of model **2** on the other hand locates on both the dirhenium atoms and the sulfide ligands. The LUMO (abs. as L) of model **2** has $d_{mm\sigma*}$ character as well as $p\sigma^*$ antibonding of oxo ligand.

Electronic Excitations

The gas-phase excitation spectra of models **1** and **2** are depicted in Fig. 3. The intense lowest-energy excitation peak (λ_{max}) of model **1** locates at 461 nm ($f = 0.01$). It mainly consists of the orbital-pair transition (H, L + 2) (99%) from the HOMO to the LUMO + 2. As mentioned, the HOMO locates on the two equivalent Re atoms with the metal–metal antibonding orbital character $d_{mm\delta*}$ and the LUMO + 2 also mainly locates at the two Re atoms but with the metal–metal π antibonding orbital character (noted as $d_{mm\pi*}$). Consequently, the λ_{max} of model **1** at 461 nm involves the $d_{mm\delta*} \rightarrow d_{mm\pi*}$ transition. The λ_{max} of model **2** locates at 546 nm which red-shifts 85 nm to that of model **1**. It has a slightly weak intensity of the calculated $f = 0.008$ and mainly composes of the orbital-pair transition of (H, L + 1). The HOMO is characterized by the p orbitals of the S atoms and the carbon atoms of L fragments (noted as L_σ) and metal–metal δ bonding (noted as $d_{mm\delta}$) of dirhenium. The LUMO + 1 mostly consists of the $5d$ orbital components of two Re atoms with $d_{mm\sigma*}$ character and a small percent of $p\sigma$ antibonding orbital components of oxo ligand ($L_{\sigma*}$). Consequently, The CT processes involved in λ_{max} transition are from (L_σ + $d_{mm\delta}$) to ($d_{mm\sigma*}$ + $L_{\sigma*}$).

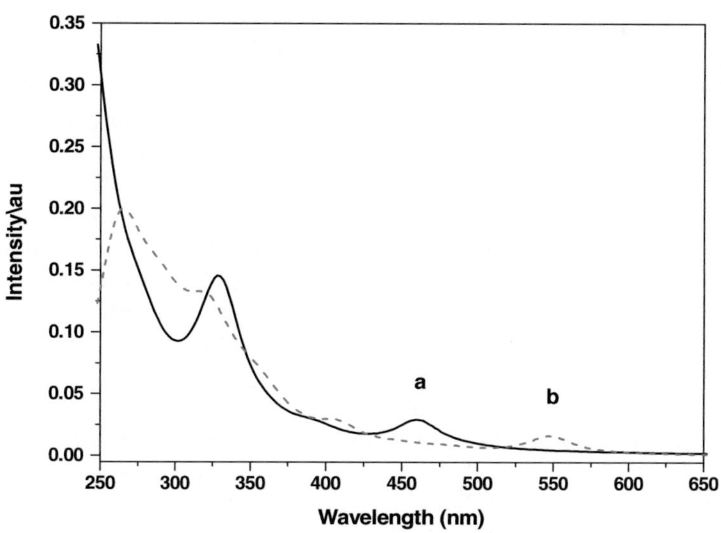

Fig. 3 Simulated electronic excitation spectra of (**a**) model **1** (*line*) and (**b**) model **2** (*gray dash*)

The measurements of the UV-vis spectra of these two cluster compounds at present are not available. Eglin and his collaborators have reported the intense lowest-energy absorption of another strong Re–Re coupled cluster [$Re_2Cl(NCN)_4$] [BF_4] (Re–Re distance is 2.224 Å which is very closed to that of model **1,** 2.225 Å) located at 460 nm with typical $d \rightarrow d$ transition character [98]. Some other available reports also give rise to the λ_{max} at around 420 nm–470 nm of some dirhenium complexes containing strong Re–Re interactions [99, 100].

Polarizabilities and Quadratic Hyperpolarizabilities

The calculated polarizability matrices of models **1** and **2** are almost diagonal, i.e., $\alpha_{ij} = 0\ (i \neq j)$. The spatially averaged $\overline{\alpha}$ value of model **1** is 33 $\times$ 10^{-24} esu which is about one half of that of model **2** (67 $\times$ 10^{-24} esu).

The anisotropy of the calculated static β_i values of both models is obvious ($\beta_z \gg \beta_x, \beta_y$) due to the specified molecular Cartesian coordinates. The quadratic hyperpolarizability of model **1** [$\beta_z(\mathbf{1}) \approx \overline{\beta}(\mathbf{1}) = 0.6 \times 10^{-31}$ esu] is about one order of magnitude smaller than that of model **2** [$\beta_z(\mathbf{2}) \approx \overline{\beta}(\mathbf{2}) = 7.0 \times 10^{-31}$ esu]. The result is a surprise in that the second-order NLO response of model **1** with strong Re–Re interaction is much smaller than that of model **2** which contains weak Re–Re interaction.

The orbital decomposition analysis scheme has been executed to reveal the contributions of the particular occupied-virtual orbital-pair transitions (*a*, *b*) to a given quadratic hyperpolarizability as well as the corresponding CT processes involved in the transitions. The relevant orbital-pair transitions of polyatomic model **1** are multicomponent. The main contribution of the first four major orbital-pair transitions in magnitude order is that (H − 1, L + 6), (H − 4, L + 3), (H − 2, L), and (H − 1, L + 1) have the positive signs, while the fifth one (H, L + 4) has the negative sign as illustrated in Fig. 4a. They are all obviously characterized by predominant metal-to-metal transitions relating to MMCT processes. The orbital pair (H − 1, L + 6) has the largest positive contribution (the relative ratio is assumed as 100%). It is involved in the relative high-energy excitations (in the range from 4.9 to 6.1 eV). We note that it is *excluded* in the optically intense excitation (λ_{max}). The result conflicts with the assumption of the traditional two-level model which assumes that only the lower energy transitions are counted. The occupied H − 1 is the δ bond orbital of dirhenium ($d_{mm\delta}$) while the virtual L + 6 is dominant by π antibond orbital of dirhenium ($d_{mm\pi^*}$) with the fractional components of $p\sigma^*$ orbitals of the allyl ligands (L_{σ^*}) as shown in Fig. 4a. This orbital-pair transition therefore involves the MMCT ($d_{mm\delta} \rightarrow d_{mm\pi^*}$) and MLCT ($d_{mm\delta} \rightarrow L_{\sigma^*}$) processes. The former process is obviously predominant (~80%). The analyses showed that the following two orbital-pair transitions of (H − 4, L + 3) and (H − 2, L) have the second and third largest contribution to the $\overline{\beta}$ with the relative ratio of 88% and 51%, respectively. Both mainly involve MMCT with $d_{mm\pi} \rightarrow d_{mmsd^*}$ and $d_{mm\pi} \rightarrow d_{mm\sigma^*}$ characters, respectively (d_{mmsd^*} denotes

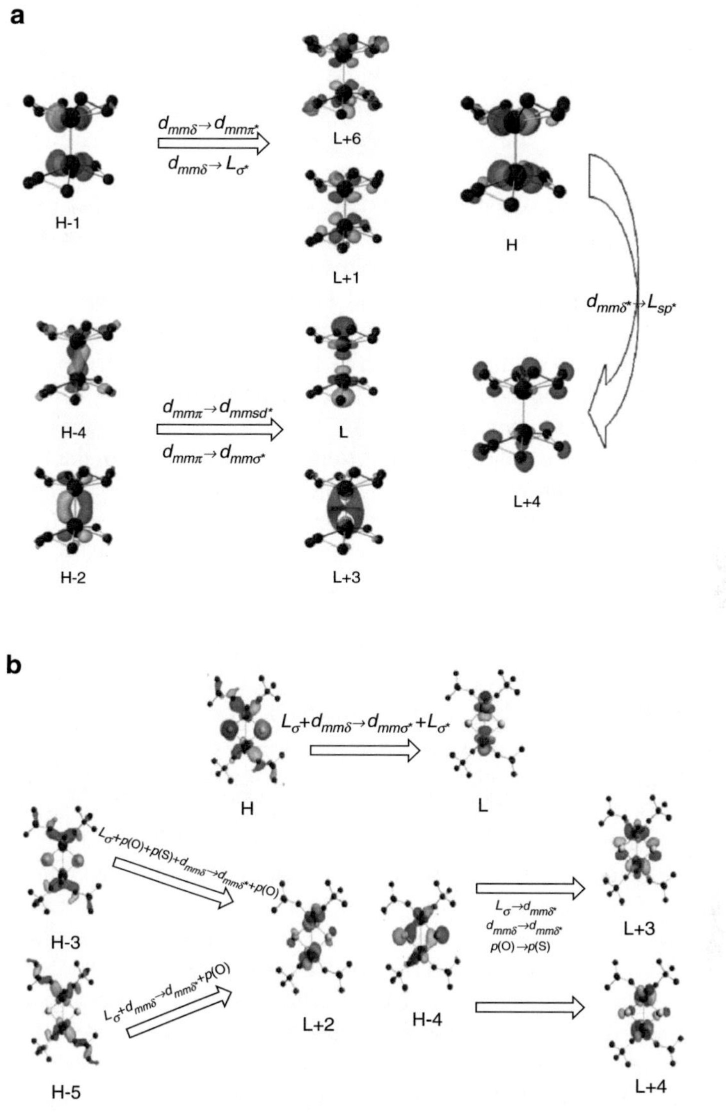

Fig. 4 Illustrations of the orbital-pair transitions relevant to the quadratic hyperpolarizability of model **1** (**a**) and **2** (**b**)

the *sd* hybrid antibonding orbital of dirhenium [101]). Another orbital pair (H − 1, L + 1) with relative ratio (23%) is similar to (H − 1, L + 6). It involves predominated MMCT ($d_{\mathrm{mm}\delta} \rightarrow d_{\mathrm{mm}\pi*}$) and fractional MLCT ($d_{\mathrm{mm}\delta} \rightarrow L_{\sigma*}$) characters. It is worth noting that there exhibits a negative contributor, the orbital pair (H, L + 4). It has a nonneglectable relative ratio of about −28% to the largest one. The HOMO locates on the dirhenium with typical δ antibond character as mentioned above while the LUMO + 4 locates mostly on the allyl ligands. This orbital-pair transition therefore mainly involves MLCT ($d_{mm\delta*} \rightarrow L_{sp*}$), where L_{sp*} denotes the sp hybrid antibonding orbitals of allyl ligands. In a word, the metal-to-metal transition processes predominate over the $\overline{\beta}$ of model **1** while the MLCT process could be unfavorable to the enhancement of the quadratic hyperpolarizability.

The decomposition scheme performed on model **2** shows that there are three orbital-pair transitions contributing to β with positive signs while there are two with negative signs. The (H, L) transition has the largest contribution (100%). It is solely involved in the first dark excitation and is again excluded from the λ_{max}-related excitation. As depicted in Fig. 4b, the HOMO of model **2** is no longer dominated by the 5*d* orbitals of dirhenium, instead it has large components of the p orbitals of S atoms and CH_2CMe_3 ligand fragments: HOMO(**2**) $\approx$ 0.48*p*(S) + 0.27*p*(L) + 0.12*d* (Re–Re). The LUMO in contrast to the HOMO is dominated by the $d_{mm\sigma*}$ orbitals of dirhenium with the nonnegligible fractional contribution from the ligands, which is mostly characterized by the atomic p antibonding orbitals of O atoms: LUMO (**2**) $\approx$ 0.78*d*(Re–Re) + 0.20*p*(O). Consequently, the orbital-pair (H, L) transition has (L_{σ} + $d_{mm\delta}$) $\rightarrow$ ($d_{mm\sigma*}$ + $L_{\sigma*}$) character. The relevant transition processes are multiple, i.e., the dominant LMCT ($L_{\sigma} \rightarrow d_{mm\sigma*}$) and the fractional MMCT ($d_{mm\delta} \rightarrow d_{mm\sigma*}$) and metal-inductive LLCT ($L_{\sigma} \rightarrow L_{\sigma*}$). The second largest contribution comes from orbital pair (H − 3, L + 2), where the HOMO − 3 involves [0.47*p* (L) + 0.23*p*(S) + 0.15*d*(Re–Re)] and the LUMO + 2 involves [0.55*d*(Re–Re) + 0.23*p*(O) + 0.18*p*(S)]. The orbital components of (CH_2CMe_3) ligands [*p*(L)] which are mainly involved in the occupied H − 3 (47%) is absent in the virtual orbital L + 2 indicating its electron donor character. The relevant electronic transition processes are again LMCT (dominant), MMCT, and LLCT. The third largest contribution comes from (H − 5, L + 2) transition. Different to H − 3 orbital, H − 5 mostly locates on CH_2CMe_3 ligand fragments and dirhenium without the component of p(O) and *p*(S). The relevant transition processes can be assigned to LMCT, MMCT, and LLCT as well. The CH_2CMe_3 ligand fragments clearly play the role as the electronic donor in the LMCT and LLCT and S and O atoms clearly play as the electronic acceptor in LLCT. The above three orbital-pair transitions make the positive contributions to the magnitude of β values. They give priority to LMCT processes but MMCT and metal-inductive LLCT processes are impressive. They are all involved in the lower energy excitations (in the range from 2.1 to 3.6 eV). On the other hand, the contribution of the orbital-pair transition (H − 4, L + 3) is the second largest in magnitude with the relative ratio of 74%, but it is negative, i.e., it would reduce rather than enhances the quadratic hyperpolarizability. This orbital-pair transition is involved in the higher energy excitations (in the range from 3.6 to 3.8 eV). The orbital HOMO − 4 has large components of *p*(O) and *p*(S). The d orbital components of dirhenium are only about 10% in total. In the virtual orbital L + 3 on the other side, the d orbital components of dirhenium enhance to 55% in total while the p(O) components decrease greatly. The relevant electronic transition processes are thus assigned to LMCT ($L_{\sigma} \rightarrow d_{mm\sigma*}$), MMCT ($d_{mm\delta} \rightarrow d_{mm\delta*}$), and LLCT [*p*(O) $\rightarrow$ *p*(S)]. It is worth noting that the contributions of the CH_2CMe_3 ligand fragments are absent. The other negative contributor is orbital pair (H − 4, L + 4). Its contribution in magnitude has a relative ratio of about 18% to the largest one. The analysis shows the similar situation to that of orbital pair (H − 4, L + 3) and will not be described in detail for clarity. It also involves the LMCT, MMCT, and LLCT processes missing the contribution of CH_2CMe_3 ligand fragments. In a word, the main CT process contributing to the quadratic hyperpolarizability of model **2** is LMCT process which is completely different to the situation in model **1**. If the CH_2CMe_3 fragments participate in the LMCT process as the electron donors in addition to the oxo and sulfide ligands, the contribution would be positive or vice versa. The MMCT has been confirmed to be one of the transition processes that play roles in the enhancement of the β values.

2.3 *Second-Order NLO Properties of Trinuclear Anionic Clusters*

In this subsection, we present the first-principle DFT calculations on the electronic excitations and second-order NLO properties in *solution phase* of two typical inorganic trinuclear anionic clusters, $[MoCu_2S_4(SPh)_2]^{2-}$ and $[Mo_2CuS_4]^{1-}(edt)_2(PPh_3)$ (*edt* = 1,2-ethanedithiolato). The computed excitation energies are in good agreement with the outcome of the measurements. The predicted values of the molecular quadratic hyperpolarizabilities are of the comparable order of those of the typical organometallic chromophores. We demonstrate again the significant contributions to the second-order responses from the charge transfers between the metal centers (MMCT) in these two charged clusters. This meaningful *ligand-independent* mechanism for the second-order response largely relates to metal–metal bonding strength.

2.3.1 Structures and Computational Details

The two molybdenum–copper trinuclear anionic clusters have been chosen due firstly to the various structures in relation to the metal *nd* configuration and oxidation states in the copper–tetrathiomolybdate complexes. One is a linear-typed bivalent anionic cluster, $[(PhS)CuS_2MoS_2Cu(SPh)]^{2-}$ (**3**), and the other is a triangular univalent anionic cluster with the incomplete cubane-like cluster core, $[Mo_2CuS_4]^{-1}(S_2C_2H_4)(PPh_3)$ (**4**) [102, 103]. The second feature lies in the similar Mo–Cu bonding in these two charged clusters possessing the different metal-core configurations (linear-typed vs. triangular, see Fig. 5). The original structures came from the X-ray diffraction data. In view of the metal–metal interaction, both clusters exhibit weak *direct* metal–metal interactions. Cluster **3** contains a pair of equivalent Mo–Cu bonds slightly shorter than the sum of atomic radii. Cluster **4** contains two inequable Mo–Cu bonds and an additional Mo(1)–Mo(2) bond. Although the metal–metal distances in cluster **4** are slightly larger than the sum

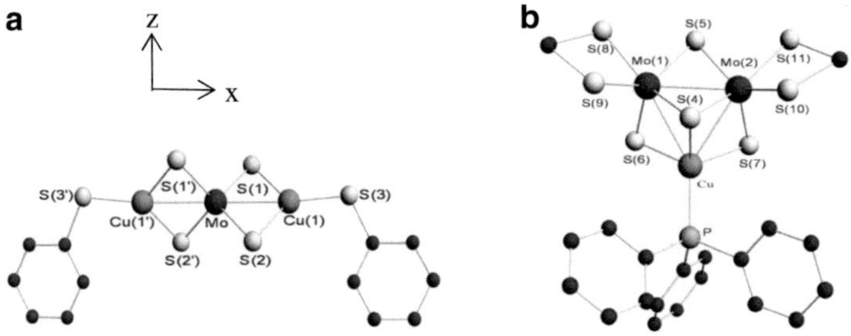

Fig. 5 Molecular structures of the anionic clusters (**a**) $[(PhS)CuS_2MoS_2Cu(SPh)]^{2-}$, (**b**) $[Mo_2CuS_4]^{1-}(S_2C_2H_4)(PPh_3)$. H atoms are omitted for clarity

of atomic radii, the *direct* weak metal–metal bonding interactions are generally considered [103].

The molecular geometries of clusters **3** and **4** were fully optimized in acetonitrile solution to the local energy minima which have been confirmed by no imaginary harmonic vibration frequency. The computed geometric parameters such as the bond lengths and angles are in reasonable agreement with the reported X-ray diffraction data. The optimizations and ground-state SCF calculations were proceeded at the TZ2P Slater-type orbital basis set with two augmented polarization functions and the "small" frozen core level: (Mo:3*d*; Cu: 2*p*; O:1*s*; S:2*p*; N:1*s*). The BP pure GGA XC functional was used with the LDA part being VWN type including the Stoll correction. The scaled-ZORA approximation of the relativistic theory was used. The solvent effects were employed in this study by using conductor-like screening model (COSMO) [104–106] of solvation with the Klamt surface [107]. The solute dielectric constant were set to 37.5 (MeCN, acetonitrile) for both models. The electronic excitation properties and the quadratic hyperpolarizabilities were calculated by using the response theory implemented in the RESPONSE module of ADF program. The GGA part of the XC functional employed the GRAC of the potentials based on the BP functional and the shape-corrected LB94 potential. Only the static β values of two models were calculated due to the absence of the comparative experimental dynamic data. Based on this consideration, the frequency-dependent effects as well as intermolecular interaction effects which have been proved to be quantitatively important are not included in the present qualitative evaluations of the quadratic hyperpolarizabilities.

2.3.2 Results and Discussion

Electronic Structures and Metal–Metal Bonding

The optimizations in solution can ensure the valid equilibrium structures for the solution-phase calculations of the electronic excitations. The small difference from the crystal structures can be explained by the presence of the intercluster interactions in solid state. Cluster **3** has C_2 point group symmetry with z axis being the twofold axis. The central molybdenum atom is fourfold coordinated by four bridged sulfur atoms [μ_2-S(1, 2)]. Each copper atom is threefold coordinated by two μ_2-S(1, 2) and one terminal sulfur atom [t-S(3)]. The fragmental core MoS(1)S(2) Cu(1) is almost planar. And the structure features that the two equivalent planar cores [MoS(1)S(2)Cu(1) and MoS(1′)S(2′)Cu(1′)] are perpendicular to each other. The π-conjugated phenyl ring (noted as Ph in the following) connects to copper via t-S(3) with an angle of $\angle$Cu-S(3)-Ph $= 111.0°$. The distance between Mo and Cu atoms (2.669 Å) is about 0.03 Å longer than the sum of atomic radii ($r_{\text{Mo–Cu}} = 2.639$ Å). Cluster **4** features a near-regular trigonal fragmental core Mo(1)Mo(2)Cu. Each Mo atom coordinates to five sulfur atoms in a tetragonal-pyramidal manner. The Mo atoms directly connected to two *edt* ligands [$S_2(CH_2)_2$]. The sole Cu atom is tetrahedrally coordinated by three sulfur atoms and one PPh_3 ligand. Cluster **4** has two inequivalent Mo–Cu bondings in contrast to cluster **3** does, and it has an additional Mo(1)–Mo(2) bond. Mo(1)–Cu and Mo(2)–Cu distances (2.792 Å and 2.807 Å, respectively) are about 0.16 Å longer than the sum of atomic radii, and Mo(1)–Mo (2) distance is 0.15 Å longer than the sum of atomic radii ($r_{\text{Mo–Mo}} = 2.726$ Å).

The frontier molecular orbitals of polyatomic clusters **3** and **4** are multicomponent. The description will be emphasized for clarity on the metal characters and the metal–metal bonding properties. The *occupied* molecular orbitals of cluster **3** are characterized by the 3d valent orbitals of Cu atoms (noted as d_{Cu}) and 3p orbital of the sulfur atoms (noted as p_{S}). The Mo character is absent in the first three occupied MOs. The p–π orbitals of phenyl rings of PPh_3 ligands (noted as π_{Ph}) are present but minor. On the contrary, the 4d valent orbitals of Mo atoms (d_{Mo}) are dominant in the first three *unoccupied* molecular orbitals. The orbital component of MoS_4 fragment in LUMO is about 80%, while d_{Cu} are also present but is minor while p_{S} [t-S(3)] and π_{Ph} are absent. The anti-π bonding (noted as $d_{\text{mm}\pi*}$) and anti-δ bonding (noted as $d_{\text{mm}\sigma*}$) can be clearly found indicating the *direct* Mo–Cu interactions. The first two occupied MOs of cluster **4** are characterized by p–π orbitals of the *edt* ligands (p_{edt}) with a small amount of d_{Cu}. The third one (HOMO − 2) on the other side is characterized by the weak Mo(1)–Mo(2) π-bonding orbital (noted as $d_{\text{mm}\pi}$) with a small component of d_{Cu}. The first three *unoccupied* MOs are dominant by the Mo(1)–Mo(2) antibonding. Mo(1)–Mo(2) anti-δ bonding (noted as $d_{\text{mm}\delta*}$) presents in LUMO, $d_{\text{mm}\pi*}$ presents in LUMO + 1, and $d_{\text{mm}\sigma*}$ presents in LUMO + 2. Since there are single bonds between Mo and μ-S atoms (S_4–S_7), it is better to analyze the character of the Mo_2S_4 fragment as a whole rather than the separate Mo atoms and μ-S atoms.

Electronic Excitations

The computed vertical singlet excitation energies of models **3** and **4** in solution phase are illustrated in Fig. 6. The excitation bands of the trinuclear clusters are multicomponent. For concise, the emphasis is put on the main characteristics of each band as well as the calculated energies in comparison with the measurements.

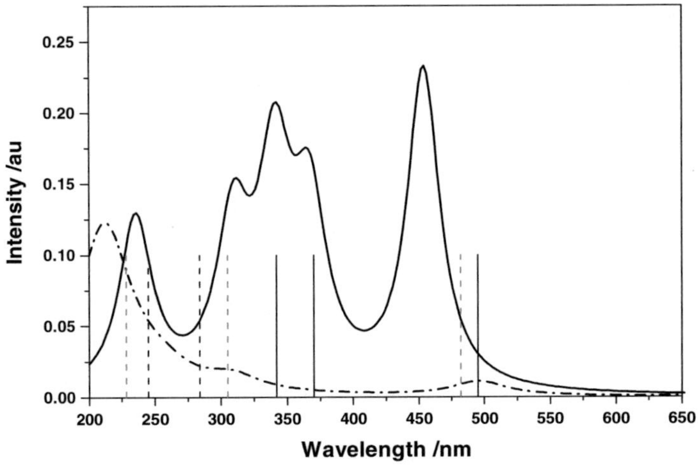

Fig. 6 Simulated electronic excitation spectra of anionic clusters **3** (line) and **4** (*dash dot*) in solution phase. The experimental excitation energies (*vertical line* for cluster **3** and *vertical dot* for cluster **4**) are also illustrated

Cluster **3** was computed to have five distinct absorption bands centering at 456 nm (ν_1), 361 nm (ν_2), 331 nm (ν_3), 312 nm (ν_4), and 235 nm (ν_5). The calculated results agreed well with the measured data obtained in acetonitrile solution, i.e., 495 nm, 370 nm (sh), 342 nm, 284 nm, and 245 nm, respectively [102]. The analyses provide detailed orbital-pair transitions involved in the excitaion bands. The lowest-energy band ν_1 (λ_{max}) contains three vertical excitations in a narrow energy range (2.68 eV–2.82 eV). It mainly originates from the charge transfers from *t*-S(3) and Ph ligand to the central Mo atom charactering the LMCT. ν_2 band contains two intense excitations with the energies of 3.42 eV and 3.43 eV. The MLCT character is dominant with the CT from [MoCuS] cores to Ph rings. ν_3 band contains three excitations with the energies from 3.68 to 3.86 eV. The main CT involved is MLCT ([MoCuS] fragment → Ph rings) while LLCT characters the minor CT process [*t*-S(3) → μ-S(1,2)]. ν_4 band contains only one excitation (3.97 eV) which agrees well with the measured data of 284 nm. It manily involves the typical MLCT ([MoCuS] → Ph). The high-energy band ν_5 contains two excitations with the energies of 5.24 eV and 5.36 eV. In addition to the typical MLCT involved in this band [CuS(1,2) → SPh], the d_{Cu}–d_{Mo} transitions are clearly presented. The MMCT (Cu → Mo) appears in the high-energy excitations.

The computed electronic excitations of cluster **4** display four intense bands. They are centered at 496 nm (ν_1), 309 nm (ν_2), 255 nm (ν_3), and 212 nm (ν_4), respectively. The three measured absorption bands are located at 482 nm, 305 nm, and 228 nm [103]. The theoretical values reasonably agreed with the experimental record except ν_3 which was not observed in the measurement. The lowest-energy band ν_1 (λ_{max}) contains single intense excitation, which is dominant by the *d–p* transitions with a small component of d_{Cu}–d_{Mo} transition. The CT processes involved in the excitation are short-distance MLCT/LMCT and weak MMCT

(Cu → Mo). v_2 band contains three vertical excitations in a narrow energy range around 4.02 eV. It characters the CT processes along the *z* axis from Mo_2S_4Cu core to Ph rings (MLCT) and *edt* to Ph rings (LLCT). v_3 contains six excitations, which are commonly characterized by the *p–p* transitions (*edt* → Ph and Ph → *edt*) and *d–p* transition (Mo_2S_4Cu → Ph). The high-enery band v_4 is realtively broad and intense containing nine singlet excitations in the energy range from 5.55 eV to 5.94 eV. It is dominant by the CT between metals (Cu → Mo) and the intra-*edt*-ligand CT [*edt* (S) → *edt*(C)]. The MMCT again appears in the high-energy excitations.

Polarizability and Quadratic Hyperpolarizabilities

The distinct structural anisotropy of two models has been sufficiently considered in assigning the Cartesian coordinates for the hyperpolarizability computations. The twofold axis of model **3** was arranged to be along *z* axis which is also identical to the dipole moment direction. The *z* axis of model **4** is along the direction of the vertical line of Mo(1)–Mo(2) bonding via Cu, and the *xz* plane contains the trigonal Mo(1) Mo(2)Cu core. As a result, the coordinate-dependent tensor components of polarizability and hyperpolarizability of both models exhibit the anisotropic characteristics. The polarizability tensor is almost diagonal ($\alpha_{ij} \approx 0,\ i \neq j$). The spatially averaged $\overline{\alpha}$ value of model **3** is 65 × 10^{-24} esu which is about two third of that of model **4** (92 × 10^{-24} esu). The static β_i values of the two models is remarkably anisotropic ($\beta_z \gg \beta_x,\ \beta_y$) resulting in $\beta_z \approx \overline{\beta}$. The quadratic hyperpolarizability of model **3** [$\beta_z(\mathbf{3}) \approx \overline{\beta}(\mathbf{3}) = 6.7 \times 10^{-30}$ esu] is about one half of that of model **4** [$\beta_z(\mathbf{4}) \approx \overline{\beta}(\mathbf{4}) = 16.0 \times 10^{-30}$ esu]. Since the absence of the experiental data of the quadratic hyperpolarizability, the computed quantities are only comparatively meaningful, i.e., model **4** possesses larger quadratic hyperpolarizability than model **3** does. And these values are modest and comparable to those of the typical organometallic NLO chromophores [15] such as that of metal carbonyl containing prindine or styrylpyridine ligands.

The contributions of the particular orbital-pair transitions (*a*, *b*) to β are unveiled by using the orbital decomposition scheme. The relevant orbital-pair transitions of model **3** are multicomponent. Figure 7 showed the dominant transitions and the their relative contributions. The (H, L + 8) transition makes the largest contribution to β_{zzz} (100%). It is mainly involved in the high-energy excitations (in the range from 4.9 eV to 6.1 eV). It is worth to note that this orbital-pair transition is again excluded from the lowest energy intense excitation (i.e., λ_{max}). The *virtual* LUMO + 8 is dominant by π^* orbitals of Ph rings ligands ($ph_{\pi*}$). This transition therefore involves the MLCT process [CuS(3) → Ph]. The second contribution came from (H − 2, L + 7) (53%), which also involves the MLCT similar to the first one. The typical $d_{Cu} \rightarrow d_{Mo}$ transition is the main character of the third contribution (21%) of (H − 9, L) transition. The MMCT (Cu → Mo) contribution to β value from this transition is impressive. The following contribution of (H − 3, L + 1) transition (14%) again exhibits the impressive MMCT character (Cu → Mo) which reinforces the MMCT contribution to the global β_{CT}. The (H − 2, L + 1) transition

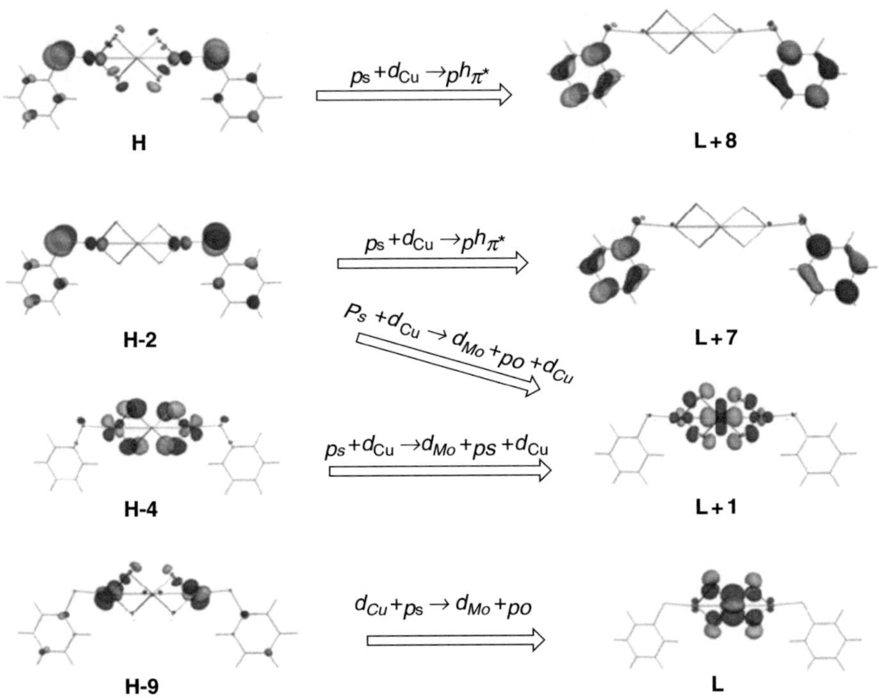

Fig. 7 Illustrations of the orbital-pair transitions and corresponding CT routes involved in the significant contribution to the quadratic hyperpolarizability of cluster **3**

has negative contribution (−19%) which is unfavorable to the global second-order response. The transition is a typical *p*–*d* transition involving the CT from SPh ligand to the MoS_4Cu core (LMCT). In summary, the multiple CT processess are responsible to the origin of the second-order response of model **3** in which the MMCT process is confirmed with the positive contribution to global β value while the LMCT from the ligands to the metal core reduces the global β value.

There are six orbital-pair transitions mainly involved in the contribution of β value of cluster **4** (Fig. 8). The first two transitions, (H, L + 1) and (H − 1, L), have the larger contribution (100% and 68%, respectively) than the other four ones. (H, L + 1) transition characters the $p_{edt} \rightarrow d_{mm\pi^*}$ transition involving the LMCT process (*edt* → Mo_2S_4). The second (H − 1, L) contribution dominates the LMCT (*edt* → Mo_2S_4) process similar to the first one. The following four orbital-pair transitions, i.e., (H − 6, L + 1), (H − 2, L + 5), (H − 5, L + 2), and (H − 5, L), have the similar mechanism, and one of them (H − 6, L + 1) is thus taken as the example. The HOMO − 6 possesses the significant metal charater of d_{Cu} and the p_{edt} [$0.22d(Cu) + 0.37p_{edt} + 0.11p(S) + 0.05p(P)$] while the *unoccupied* LUMO + 1 consists of a large component of $d_{Mo\text{–}Mo}$ with $d_{mm\pi^*}$ character [$0.40d_{mm\pi^*} + 0.06d(Cu) + 0.08p_{edt} + 0.27p(S)$] with the greatly reduced d_{Cu} and p_{edt} components. This transition involves the LMCT (*edt* → Mo_2S_4) and MMCT

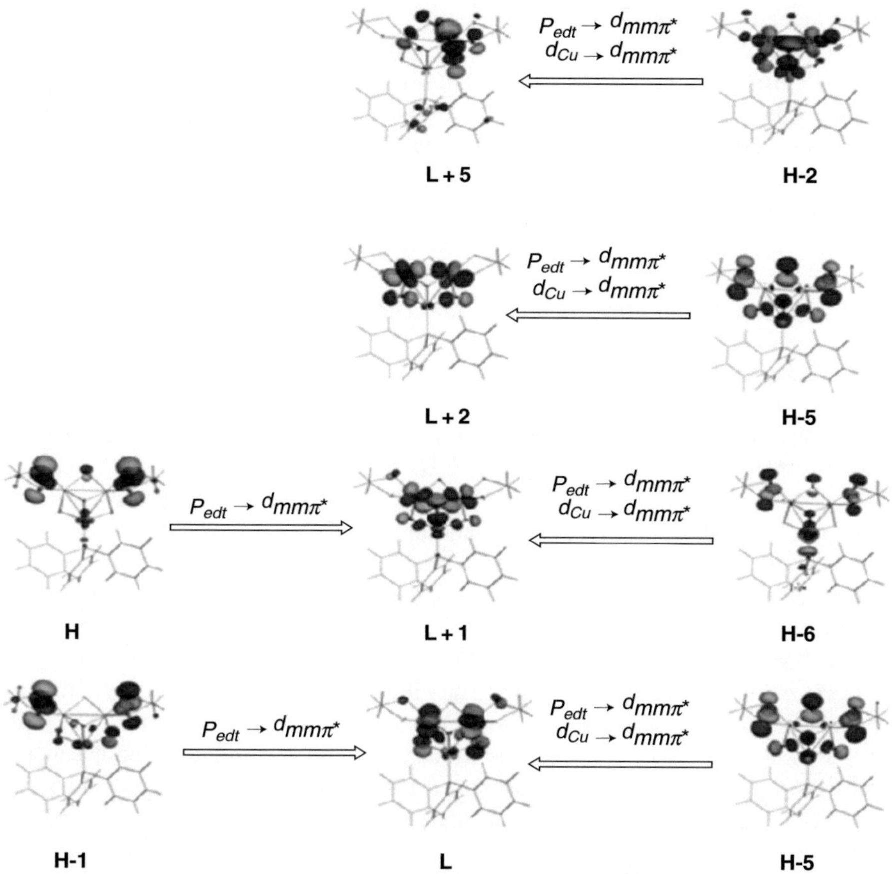

Fig. 8 Illustrations of the orbital-pair transitions and corresponding CT routes involved in the significant contribution to the quadratic hyperpolarizability of cluster **4**

($d_{Cu} \rightarrow d_{mm\pi^*}$, Cu $\rightarrow$ Mo_2S_4), respectively. The later one is *nonnegligible* which once again exhibits the MMCT contribution to the second-order activity.

The MMCT processes in both models are significant for both models. The *ligand-independent* MMCT process is responsible to the fractional amount of global β_{CT} value particular in the case of model **4** of about 20%.

2.4 Second-Order NLO Properties of Pentanuclear Clusters

We present in this subsection the second-order nonlinear optical properties of a series of pentanuclear metal clusters $[MS_4Cu_4X_2Py_6]$ (M = Mo, W; X = Br, I) on the basis of the hyper-Rayleigh scattering (HRS) experiments and the first-principle

DFT calculations. The measurements obtain the notably large dynamic quadratic hyperpolarizabilities at 1,064 nm [$\beta(-2\omega, \omega, \omega)$ values are around 200 × 10^{-30} esu] and, by extrapolation, a large static values around 60 × 10^{-30} esu. The computational results of the electronic excitation energies and quadratic hyperpolarizabilities are in good agreement with the experimental. The in-depth analysis of the mechanism for the second-order response unambiguously shows the evidence of the contribution of *direct* metal–metal interaction charge transfers.

2.4.1 Experimental and Computational Details

HRS Measurement

The molecular quadratic hyperpolarizabilities of the four pentanuclear clusters, namely, [$MS_4Cu_4X_2Py_6$] (M = W, X = Br for **5** and I for **6**; M = Mo; X = Br for **7** and I for **8**) have been determined by HRS technique [108, 109]. The external reference method was utilized in the measurements by choosing *para*-nitroaniline (*p*NA) as standard. The measurements were carried out in dimethylformamide (DMF) solutions at 1,064 nm fundamental wavelength pumped from the optical parameter oscillation (OPO) by the Q-switched Nd:YAG laser (10 Hz, 8 ns pulse width). The laser radiation was focused by a cylindric lens (focal length 10 cm) into a quartz cell containing the samples. The HRS signals were detected by a fluorescence spectrometer containing a monochromator and a photo counter with high-degree of accuracy. The weak contributions of two-photon-induced fluorescence in the measurements were directly subtracted from total signal intensity to obtain the pure HRS signal by using a high-resolution monochromator near 532 nm. The solutions were sufficiently diluted (the condensations were below 4 × 10^{-4}) to ensure that absorption of scattered second-harmonic light was negligible.

The HRS results of the quadratic hyperpolarizabilities are frequency dependent. The corresponding inherent static values are extrapolated by using the two-level formulation [110]

$$\beta^0_{\mathrm{HRS}} = \beta^{\lambda}_{\mathrm{HRS}}\left[1 - \left(\frac{2\lambda_{\max}}{\lambda}\right)^2\right]\left[1 - \left(\frac{\lambda_{\max}}{\lambda}\right)^2\right] \tag{13}$$

where $\lambda_{\max}$ are the absorption maximum wavelengths of compounds and λ is laser radiation wavelength.

Structure and Computational Methods

The molecular geometries were fully optimized in DMF solution modeled by COSMO approach using the Klamt surface. The solute dielectric constant was 36.0 for DMF. The TZP Slater-type basis set was used with the frozen core scheme of (W:4*f*; Cu:3*p*; S: 2*p*; N:1*s*). The BP XC functional was used with LDA part being VWN type including Stoll correction. The scaled-ZORA Hamiltonian was used to take account of the relativistic effect. The electronic excitations and the quadratic hyperpolarizabilities have been calculated in DMF solution by using GRAC potentials based on the shape-corrected LB94 potential. The Davidson diagonalization method has been used to yield the excitation information.

The HRS comparable values β_{HRS} are also calculated by using the formula deduced by Cyvin et al. [111] assuming Kleinman's symmetry and plane-polarized incident light:

$$\beta_{HRS} = \sqrt{\langle \beta_{ZZZ}^2 \rangle + \langle \beta_{XZZ}^2 \rangle} \tag{14}$$

where

$$\langle \beta_{ZZZ}^2 \rangle = \frac{1}{7}\sum_i \beta_{iii}^2 + \frac{6}{35}\sum_{i \neq j} \beta_{iii}\beta_{ijj} + \frac{9}{35}\sum_{i \neq j} \beta_{iij}^2 + \frac{6}{35}\sum_{ijk,\ \text{cyclic}} \beta_{iij}\beta_{jkk} + \frac{12}{35}\beta_{ijk}^2$$

$$\langle \beta_{XZZ}^2 \rangle = \frac{1}{35}\sum_i \beta_{iii}^2 - \frac{2}{105}\sum_{i \neq j} \beta_{iii}\beta_{ijj} + \frac{11}{105}\sum_{i \neq j} \beta_{iij}^2 - \frac{2}{105}\sum_{ijk,\ \text{cyclic}} \beta_{iij}\beta_{jkk} + \frac{8}{35}\beta_{ijk}^2$$

The full expression for β_{HRS} can be found in the reference [112].

2.4.2 Results and Discussion

HRS Results

The HRS results of the quadratic hyperpolarizabilities are summarized in Table 3. The two bromide clusters **5** and **7** have slightly different quadratic hyperpolarizabilities (β_{HRS}). Cluster **7** which is centered by Mo atom has a larger β_{HRS} than cluster **5** does which has W as the central metal atom. The condition is similar to the two iodate clusters. The β_{HRS}(**8**) containing Mo central atom is about 20% larger than β_{HRS}(**6**). On the other hand, clusters **6** and **8** coordinated to iodine atoms have much larger β_{HRS} values than clusters **5** and **7** does which are coordinated to

Table 3 HRS experimental results of molecular quadratic hyperpolarizability (10^{-30} esu) in DMF solution of clusters **5**–**8** at 1,064 nm laser radiation

Clusters	β_{HRS}
$[WS_4Cu_4Br_2Py_6]$ (**5**)	164
$[WS_4Cu_4I_2Py_6]$ (**6**)	319
$[MoS_4Cu_4Br_2Py_6]$ (**7**)	198
$[MoS_4Cu_4I_2Py_6]$ (**8**)	354

bromine atoms. For example, the β_{HRS}(**6**) value is about twice as β_{HRS}(**5**). This is understandable because of the larger ionic radius of iodine atom, although both iodine and bromine have the same number of valent p electrons. The lone-pair electrons in iodine ionic ligands could be more delocalized in the clusters which benefit to the electron donating. Conclusively, cluster **8** which contains both the iodine ligands and central Mo metal atom has the largest value of β_{HRS} among the four analogs, which agrees with HRS measurements. The β_{HRS}(**8**) value of 354×10^{-30} esu is sizable, and it is about one-order magnitude larger than the typical tungsten carbonyl organometallic chromopores such as $W(CO)_5$(pyridine) [113].

Electronic Structures and Metal–Metal Bonding

The optimized molecular structures of the **5**–**8** clusters (Fig. 9) were all in agreement with the reported experimental data in solid state [114–116]. The optimizations performed in DMF solution simulate the experimental environments of both the UV-vis spectra and HRS measurements. The distances between the central metal M (M = Mo, W) and coppers (Cu_1/Cu_2) relaxed in the range of 0.04 Å for M–Cu_1 and 0.06 Å for M–Cu_2, respectively, after the restricted optimization. The M–Cu lengths are closed to those in [$MoOS_3Cu_3Cl(PPh_3)_3$] [2.705 Å–2.740 Å] which is reported to have weak Mo–Cu interactions [[117]]. The bond angles of Cu–M–Cu and Cu–M–Cu tend to be 90° and 180°, respectively, resulting in the quasiregular quadrangles of the planar MCu_4 cores. The optimized distances of M–S are slightly lengthened as well in the range of 0.03 Å to 0.05 Å. The overall pictures of the DFT optimized structures of the clusters depict the more compact clusters in general with the slightly relaxed metal cores. Table 4 listed the selected Mayer bond orders of the four clusters. It is interesting to see that the bond order of Mo(W)–Cu are around 0.3 to 0.4 indicating the weak *direct* M–Cu interactions in the clusters while the single

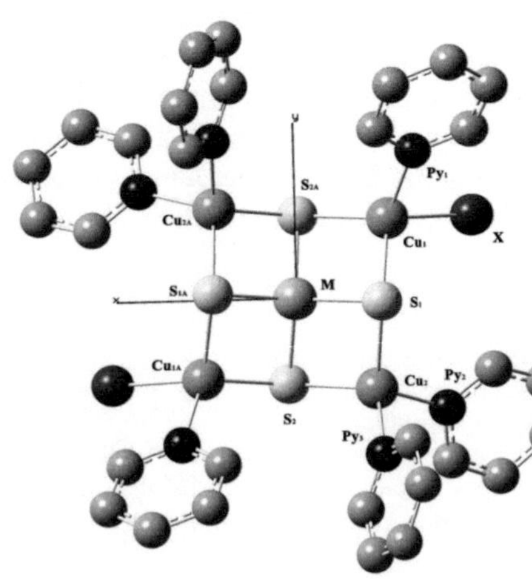

Fig. 9 Molecular structures and orientation of [$MS_4Cu_4X_2Py_6$] clusters, M = W, X = Br for cluster **5**; X = I for cluster **6**; M = Mo, X = Br for cluster **7**; and X = I for cluster **8**. H atoms are omitted for clarity

Table 4 Selected Mayer bond orders of clusters **5–7**

	5	**6**	**7**
M–Cu	0.4	0.4	0.4
M–S	1.1	1.1	1.1
Cu–S	0.6	0.6	0.7
Cu–Br/I	0.4	0.6	0.3
Cu–N(Py)	0.5	0.5	0.6

bonds between Mo(W) and S are clearly showed. The bond orders of Cu and their peripheral ligands (S, Br/I, and Py) are all around 0.5 in the range of coordination interactions.

The occupied frontier molecular orbitals of clusters **5–7** are all essentially localized on $Cu_1S_1XN(Py_1)$ and $Cu_2S_2N(Py_2)N(Py_3)$ fragments and their symmetric equivalents. The orbital overlaps between coppers and halogens are minor indicating the ionic interaction between them. The p-type orbitals of lone-pair electrons of the halogen are significant. The unoccupied MO exhibits the σ antibond orbitals of MS_4 core fragments. The LUMO of clusters **5** and **6** composes MS_4 σ^* orbitals, while for cluster **7**, both LUMO and LUMO + 1 contain this type of orbitals. The higher unoccupied MOs are essentially anti-π-conjugated MOs of the ancillary pyridine rings with only small components of MS_4 fragments. For clarity and simplicity, we analyze in detail the molecular orbitals of cluster **5**. The HOMO locates entirely on the $Cu_1S_1BrN(Py_1)$ and $Cu_2S_2N(Py_2)N(Py_3)$ and their symmetric equivalent units. The $3p_x$ orbitals of μ_3-S make σ bonds to $3d$ of Cu. The $4p_x$ of Br make *nonbond* to $3d$ of Cu_1. The HOMO is obviously irrelevant to the center metal W and Py rings. The LUMO resides mostly on WS_4 unit. The $3p$ (μ_3-S) make σ^* bonds to the $5d_{z2}$(W) atom. The LUMO + 1 mainly locates on the WS_4 core as well. The $3p(\mu_3$-S) make π^* bonds to the $5d_{xy}$(W). The situation for the other two clusters is similar.

Electronic Excitations

Table 5 listed the computed electronic excitations of clusters **5–7** in the range from 350 nm to 700 nm. The calculated excitation energies are in good agreement with the experimental outcome [114–116]. Cluster **5** has been measured to have two intensive absorptions at 442 nm and 334 nm, respectively. The calculated electronic excitation at 477 nm has 35 nm deviation for the former one, and the calculated 388 nm and 355 nm bands are close to each other with an averaged value of 371 nm which relates to the later one with a deviation of 36 nm. For clusters **6** and **7**, the deviations from the measured data are all within 40 nm.

The transitions composed in each intense excitation band are multiple and complex. Taking cluster **5** as an example, the lowest-energy exciation at 477 nm mainly involved three orbital-pair transitions of the (H − 7, L + 7), (H − 7,

Table 5 Most intensive excitations and oscillator strengths, *f*, of clusters **5–7** within the range between 350 nm and 750 nm

Cluster	Excitation	*f*	Expt.
5	477 nm	0.04	442 nm
	388 nm	0.03	334 nm
	355 nm	0.03	
6	478 nm	0.03	438 nm
	354 nm	0.09	316 nm
7	496 nm	0.03	526 nm
	355 nm	0.05	384 nm
	327 nm	0.03	298 nm

L + 5), and (H − 8, L + 4). The first transition is corresponding to the transition from 4*p* orbitals of two Br atoms to π orbitals of pyridine rings, which belongs to LLCT process. The second one is similar to the first one which belongs to LLCT as well. The third one is corresponding to the transition from the HOMO − 8, which is mainly localized on two Br atoms with small contributions of 4*d* orbitals of Cu_1/Cu_{1A} atoms to the LUMO + 4, which is mainly located on the Py rings. It belongs to LLCT (major) and MLCT (minor) process. Consequently, the lowest-energy absorption at 477 nm of cluster **5** is assigned to the LLCT (from p of LP to *p*–π of Py) and MLCT (from d of Cu_1 to *p*–π of Py) processes. Similarly, the lowest-energy excitations of clusters **6** and **7** have been assigned to LLCT (major) and MLCT (minor) processes as well. If the second-order response of these metal clusters was analyzed by the traditional two-level model which states that the β_{CT} is contributed by the intense lowest-enegy excitation, it could be ascribed to the LLCT (major) and MLCT (minor) processes missing the contribution of weak *direct* metal–metal interactions. However, the orbital-decomposition analysis gives rise to the quite different pictures of the mechanism for the second-order response of these metal clusters.

Polarizabilities and Quadratic Hyperpolarizabilities

The calculated dipole moments μ of clusters **5–7** are large and distinctively anisotropic. The dominant dipole components are along the *z* axis which is also the molecular twofold axis (crystalline *c* axis). The static polarizability matrices are almost diagonal. The average α values of clusters **5–7** are in the order of $\overline{\alpha}(\mathbf{2}) > \overline{\alpha}(\mathbf{3}) > \overline{\alpha}(\mathbf{1})$. Although all β tensor components have been calculated, Table 6 listed the representative quantities: the static spatial-averaged value $\overline{\beta}_0$ and β_{HRS} which could be comparable to the measured HRS data. The calculated β_{HRS} values are in good agreement with the extrapolated measured data. The order of the values in magnitude also reproduces the experiment: β_{HRS} (**6**) > β_{HRS} (**7**) > β_{HRS} (**5**).

The orbital-decomposition analysis on cluster **5** revealed that largest contribution (100%) to β was (H, L + 3) orbital-pair transition, which mainly came from the

Table 6 Dipole moments (debye), dipole static polarizabilities (10^{-24} esu), and static quadratic hyperpolarizabilities (10^{-30} esu) of clusters **5–7**

Cluster	μ	$\overline{\alpha}_0$	$\overline{\beta}_0 \sim \beta_z$	β_0 (HRS)	Expt.
5	18.2	101	25	40	42
6	18.8	108	41	62	75
7	18.3	103	33	46	58

excitations in the energy range from 1.95 eV to 2.00 eV. The second large contribution in magnitude came from the (H, L) transition (about 40% to the largest one). These two β_{CT}-related transitions have the lower excited energies and intensities than the intense absorption located at 477 nm (2.60 eV). However, these dark absorptions made major contributions to the quadratic hyperpolarizability. The LUMO + 3 is located on the Py_2 rings with small percentage on the WS_4 core, and HOMO and LUMO have been discussed. The CT processes related to the two orbital-pair transitions were thus assigned to the processes from the collective *d–p* system and the lone-pair p electrons to the planar coordination *p–π* ring ligands and *d–p* WS_4 core. The multiple CT processes are involved, and the synergic effects are obvious. The percent components of the HOMO, LUMO, and LUMO + 3 are approximated as 55%Cu(3*d*) + 20%S(3*p*) + 15%Br(4*p*) + 3%N (2*p*), 70%Py(π) + 10%W(5*d*) + 10%S(3*p*), and 50%W(5*d*) + 40%S(3*p*), respectively. These two transitions involved the MLCT [Cu(3*d*) → Py(π)], MMCT [Cu (3*d*) → W(5*d*)], and LLCT [Br(4*p*) → Py(π)] (see Fig. 10). The contribution of the weak W–Cu interaction to β-related CT is impressive in which the coppers play as electron donors and central tungsten as an acceptor. The later (H, L) transition, which is MMCT dominant, makes a positive global contribution of about 30% to the β_z value.

Cluster **6** presents similar results as cluster **5** but are more complex (Fig. 11). The dominant orbital-pair transition was again (H, L + 3). The second large contribution in magnitude came also from the (H, L) transition (about 85% to the largest one). The third large contribution in magitude came from the (H − 5, L + 3) transition with a relative contribution of about 40% to the largest one. The CT processes related to the first two dominant transitions are similar to the case of cluster **5** (from the collective *d–p* system of [$Cu_4S_4N_6$] unit and the LP p of iodine atoms to the *p–π* of Py ligands and the *d–p* of WS_4 core with MLCT, MMCT, and LLCT characters). The contributions of W–Cu weak interaction to the relevant CT are again impressive. The third transition involved the occupied HOMO − 5, which composed mainly of the p lone-pair orbital of iodine atoms and d orbitals of two Cu_1 atoms, and the virtual LUMO + 3 located on two Py_2 rings, and thus, the CT character was mainly the LLCT and MLCT.

For cluster **7**, the orbital-pair transitions that make largest contribution were the (H, L + 1) transition and the (H − 3, L + 2) transition, which gave rise to the second large one (about 11% to the largest one). The LUMO + 1 is mainly located on MoS_4 metal core. The CT processes related to the dominant transition were from the *d–p* system of [$Cu_4S_4BrN_6(Py)$] fragment to the *d–p* of MoS_4 core with the

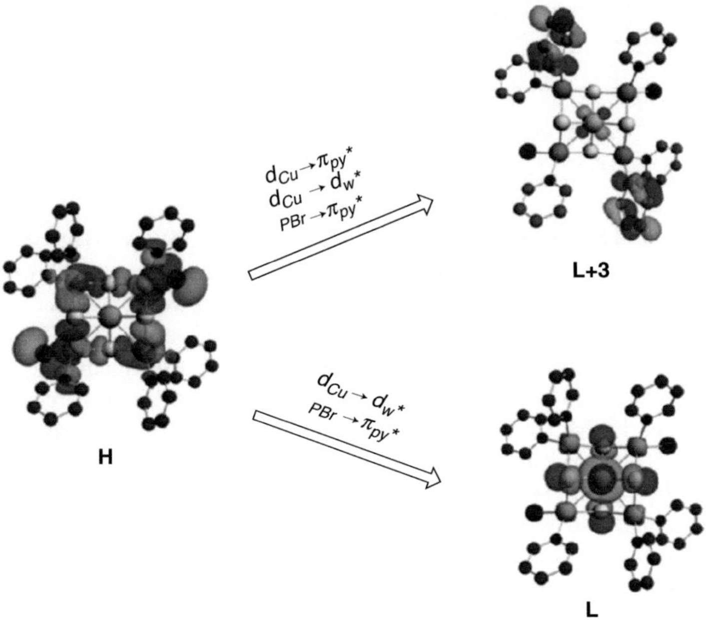

Fig. 10 Illustrations of the major orbital-pair transitions and corresponding CT routes involved in the significant contribution to the second-order responses of cluster **5**

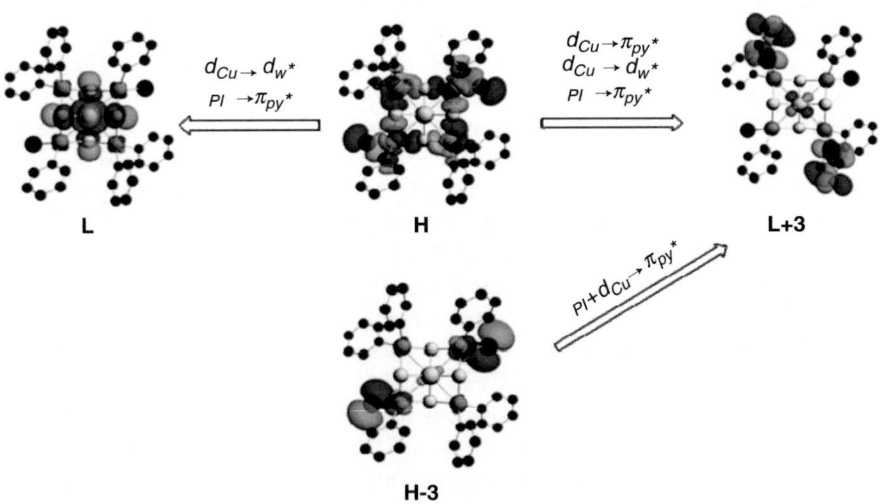

Fig. 11 Illustrations of the major orbital-pair transitions and corresponding CT routes involved in the significant contribution to the second-order responses of cluster **6**

significant character of MMCT and LLCT. The second transtion involved the charge tranfer from two $Cu_1S_1BrN(Py_1)$ fragments (HOMO − 3) to two Py_2 rings (LUMO + 2) shown in Fig. 12, and it was assigned to the typical MLCT [Cu(3*d*) → Py(π)] and LLCT [Br(4*p*) → Py(π), S(3*p*) → Py(π)] processes. Since the first transition is dominant, the contribution of Mo–Cu weak interaction CT is significant in cluster **7**.

In summary, the orbital-pair decomposition analysis gave rise to the multiple routes of the β-related CT processes and the cooperative effects contributed to the quadratic hyperpolarizabilities. The metal–ligand interactions and the metal-induced ligand–ligand interactions play the important roles in the global second-order activities similar to the cases of the organometallic chromophores. The halogen atoms with lone-pair p electrons played the roles as the electron donors in the relevant LLCT processes, while the π-conjugated pyridine ligands played the acceptors in the relevant MLCT and LLCT processes. However, the weak *direct* metal–metal interaction unambiguously involved in the β-determined CT process and in some cases even made the significant contribution (e.g., in cluster **7**), where the coppers act as the electron donors while the central metal atom (W/Mo) played the role of electron acceptor. The 3*d*(Cu) → 5*d*(W) in the case of cluster **5** exhibiting the featured d_{Cu}–d_W transitions in the polynuclear metal clusters.

The large molecular β of these clusters are favorable to produce the large macroscopic NLO coefficients. The estimation of the second-order susceptibility $\chi^{(2)}_{zzz}$ of one of the cluster compounds (cluster **7**) in solid state in terms of the molecular quadratic hyperpolarizability has been performed on the basis of the oriented-gas approximation [118] with the fact that the *z* components of the dipole moment $\mu = \mu_z$ of the molecular clusters in a unit cell have an identical direction ($\theta = \phi = \varphi = 0$).

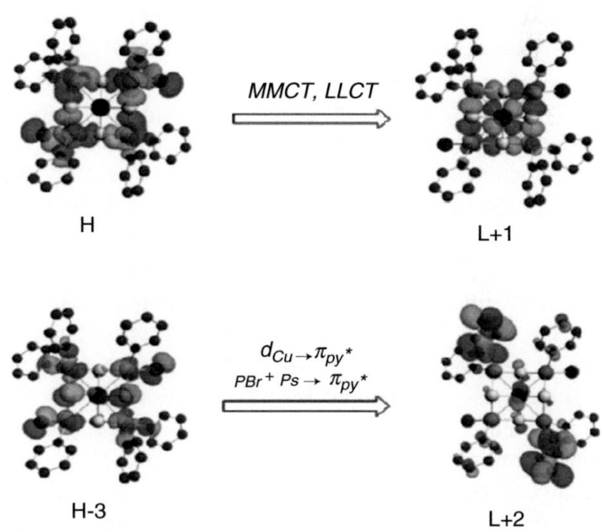

Fig. 12 Illustrations of the major orbital-pair transitions and corresponding CT routes involved in the significant contribution to the second-order responses of cluster **7**

$$\overline{\chi}^{(2)}_{zzz}(\omega) \approx NF^3\beta_{zzz}(\omega)\cos\theta\cos\phi\cos\varphi \tag{15}$$

where F is the local field factor and N is the molecular density ($N = 1 \times 10^{21}$ cm^{-3} from the experimental data). F is assumed to 1 because the measured refraction indices of the crystals are not available at present. The HRS measured β value [$\beta_{\mathrm{HRS}}(\omega) = 319 \times 10^{-30}$ esu of cluster **7**] is roughly treated as β_{zzz}. The average $\chi^{(2)}_{zzz}(\omega)$ is estimated as large as 300×10^{-9} esu which is about 10 times larger than the second-order nonlinear optical coefficients of a traditional IR nonlinear optical crystal, $AgGaS_2$ ($d_{36}(AgGaS_2) \approx 11$ pm/V ($\approx 26 \times 10^{-9}$ esu) [8]).

3 Conclusions and Perspectives

In this minireview article, we provided the recent studies on the second-order NLO properties of dinuclear, trinuclear, and pentanuclear metal cluster compounds with the comprehensive discussions on the intrinsic second-order NLO mechanism.

There exist multiple electronic transition processes that collectively contribute to the quadratic hyperpolarizability including MLCT, LMCT, MMCT, and metal-inductive LLCT. The MMCT features the NLO response mechanism of these polynuclear clusters with direct metal–metal interactions. The multiple transition processes are either cooperative to each other, such as the MLCT and MMCT synergetically enhance the β value of dinuclear model **1**, or are destructively interfering with each other, such as LMCT and MMCT can make opposite contribution to the β value of dinuclear model **2.** The traditional two-level model seems no longer valid for the complex polynuclear metal clusters because multiple excited states are coupled to the ground state to contribute to the β. For a clear description, the global β value (β_{tot}) can be denoted by the summation of the fractional β values contributed by the various CT processes: $\beta_{\mathrm{tot}} = \beta_{\mathrm{MLCT}} + \beta_{\mathrm{LMCT}} + \beta_{\mathrm{MMCT}} + \beta_{\mathrm{LLCT}} + \ldots$. For example, for anionic cluster **3**, the global β value can be accordingly denoted by $\beta_{\mathrm{tot}}(\mathbf{3}) \propto |\beta_{\mathrm{MLCT}}| - |\beta_{\mathrm{LMCT}}| + |\beta_{\mathrm{MMCT}}|$ in the magnitude order. The relatively weaker Mo–Cu interaction in model **3** results in the smaller β_{MMCT} fraction in the global β value. On the other side, the relatively intense Mo–Mo interaction in model **4** corresponds to the relatively large β_{MMCT} value ($\beta_{\mathrm{MMCT}} \sim 0.2\beta_{\mathrm{tot}}$). The intense direct metal–metal bonding interactions would result in the greater β_{MMCT} fractions in the global β_{tot} values, e.g., for a dirhenium model **1** with Re≡Re triple bond, β_{MMCT} fraction is about three fifth of the β_{tot} value.

The direct metal-to-metal transition process has been unambiguously confirmed to be the one of the second-order response mechanisms of these metal cluster compounds. This unique MMCT process could positively enhance the quadratic hyperpolarizability, and in some cases, its contribution is significant. It provides one promising means to tune the second-order NLO effects of metal cluster compounds by adjusting the metal–metal interactions. In contrast to tune the size and degree of delocalized π-conjugated ligands and the strength of the molecular dipole moments, this adjustment will less affect the IR absorptions, especially in the mid-IR region. As a result, the IR transparency (seriously depends on coordinate ligands) and nonlinearity (can rely on metal–metal bonding) can be separately tuned benefiting to the improvement of the transparency–nonlinearity trade-off. This understanding of the detailed relationship between the direct metal–metal interaction and second-order optical nonlinearity will open a way to the further investigation in the metal cluster compounds for the novel IR second-order NLO materials and the optical molecular devices.

Acknowledgments We acknowledge the financial supports from NSFC project (20973174 and 91122015) and MOST projects (2006DFA43020 and 2007CB815307).

References

1. Chai BHT (1995) Optical materials. In: Weber MJ (ed) CRC handbook of laser science and technology, vol 2. CRC, Boca Raton
2. Chemla DS (1971) Opt Commun 3:29
3. Boyd GD, Kasper H, McFee JH (1971) IEEE J QE-7:563
4. Shay LJ, Wernick JH (eds) (1975) Ternary chalcopyrite semiconductors. Growth, electronic properties and applications. Pergamon, New York
5. Schunemann PG (2006) Proc SPIE 6103:610303
6. Swain JE, Stokowski SE, Milam D, Kennedy GC (1982) Appl Phys Lett 41:12
7. Boyd GD, Buehler E, Storz FG (1971) Appl Phys Lett 18:301
8. Roberts DA (1992) IEEE J 28:2057
9. Mason PD, Jackson DJ, Gorton EK (1994) Opt Commun 110:163
10. Verozubova GA, Okunev AO, Gribenyukov AI, Trofimiv AYu, Trukhanov EM, Kolesnikov AV (2010) J Cryst Growth 312:1122
11. Byer RL, Choy MM, Herbst RL, Chemla DS, Feigelso RS (1974) Appl Phys Lett 24:65
12. Kildal H, Iseler GW (1976) Appl Opt 15:3062
13. Wu K, Chen C (1992) Appl Phys A54:209
14. Zhang G, Qin J, Liu T, Li Y, Wu Y, Chen C (2009) App Phys Lett 95:261104
15. Di Bella S (2001) Chem Soc Rev 30:355
16. Chemla DS, Zyss J (eds) (1987) Nonlinear optical properties of organic molecules and crystals. Academic, New York
17. Nalwa HS, Miyata S (eds) (1997) Nonlinear optics of organic molecules and polymers. CRC, New York
18. Marder SR, Gorman CB, Meyers F, Perry JW, Bourhill G, Bredas J-L, Pierce BM (1994) Science 265:632
19. Wolff JJ, Langle D, Hillenbrand D, Wortmann R, Matschiner R, Glania C, Kramer P (1997) Adv Mater 9:138
20. Lacroix PG, Malfant I, Iftime G, Razus AC, Nakatani K, Delaire JA (2000) Chem Eur J 6:2599
21. Brunel J, Ledoux I, Zyss J, Blanchard-Desce BM (2001) Chem Commun 923
22. Brunel J, Mongin O, Jutand A, Ledoux I, Zyss J, Blanchard-Desce M (2003) Chem Mater 15:4139
23. Goodson TG (2005) Acc Chem Res 38:99
24. Hennrich G, Murillo MT, Prados P, Al-Saraierh H, El-Dali A, Thompson DW, Collins J, Georghiou PE, Teshome A, Asselberghs I, Clays K (2007) Chem Eur J 13:7753

25. Frasier CC, Harvey MA, Cokerham MP, Hand HM, Chauchard EA, Lee CH (1986) J Phys Chem 90:5703
26. Green ML, Marder SR, Thompson ME, Bandy JA, Bloor D, Kolinsky PV, Jones RJ (1987) Nature 330:360
27. Nalwa HS (1991) Appl Organomet Chem 5:349
28. Tessore F, Roberto D, Ugo R, Quici PS, Ledoux-Rak I, Zyss J (2003) Angew Chem Int Ed 42:456
29. Powell CE, Humphery MG (2004) Coord Chem Rev 248:725
30. Cariati E, Pizzotti M, Roberto D, Tessore F, Ugo R (2006) Coord Chem Rev 250:1210
31. Coe BJ (2006) Acc Chem Res 39:383, and the references therein
32. Morrall JPL, Humphrey MG, Dalton GT, Cifuentes MP, Samoc M (2006) In: Padadopoulos MG, Sadlej AJ, Leszczynski J (eds) Nonlinear optical properties of matter, from molecules to condensed phases. Springer, Dordrecht, p 537, and the references therein
33. Senge MO, Fazekas M, Notaras EGA, Blau WJ, Zawadzka M, Locos OB, Ni Mhuircheartaigh EM (2007) Adv Mater 19:2737
34. Duncan TV, Song K, Hung S-T, Miloradovic I, Nayak A, Persoons A, Verbiest T, Therien MJ, Clays K (2008) Angew Chem Int Ed 47:2978
35. Kanis DR, Ratner MA, Marks TJ (1992) J Am Chem Soc 114:10338
36. Kanis DR, Lacroix PG, Ratner MA, Tobin JM (1994) J Am Chem Soc 116:10089
37. Matsuzawa N, Seto J, Dixon DA (1997) J Phys Chem A 101:9391
38. Karton A, Iron MA, van der Boom ME, Martin JML (2005) J Phys Chem A 109:5454
39. Asselberghs I, Therien MJ, Coe BJ, McCleverty JA, Clays K (2006) In: Schubert US, Newkome GR (eds) Metal-containing and metallosupramolecular polymers and materials. ACS symposium series, vol 928. American Chemical Society, Distributed by Oxford University Press, Washington, p 527
40. Coe BJ (2006) In: Padadopoulos MG, Sadlej AJ, Leszczynski J (eds) Nonlinear Optical Properties of Matter, From Molecules to Condensed Phases. Springer, Dordrecht, p 571, and the references therein
41. Sophy KB, Calaminici P, Pal S (2007) J Chem Theory Comput 3:716
42. Laidlaw WM, Denning RG, Verbiest T, Chauchard E, Persoons A (1993) Nature 363:58
43. Lin CS, Wu KC, Snijders JG, Sa RJ, Chen XH (2002) Acta Chim Sinica 60:664
44. de Angelis F, Fantacci S, Sgamelotti A, Cariati F, Roberto D, Tessore F, Ugo R (2006) Dalton Trans 852
45. Wu K (2006) In: Maroulis G (ed) Atoms, Molecules and Clusters in Electric Fields. Imperial College Press, London, p 565
46. Liao Y, Eichinger BE, Firestone KA, Haller M, Luo J, Kaminsky W, Benedict JB, Reid PJ, Jen AK-Y, Dalton LR, Robinson BH (2005) J Am Chem Soc 127:2758
47. Coe BJ, Harris JA, Brunschwig BS, Asselberghs I, Clays K, Garin JJ, Orduna J (2005) J Am Chem Soc 127:13399
48. Roberto D, Ugo R, Tessore F, Lucenti E, Quici S, Vezza S, Fantucci PC, Invernizzi I, Brum S, Ledoux-Rak I, Zyss J (2002) Organometallic 21:161
49. Senechal K, Maury O, Le Bozec H, Ledoux I, Zyss J (2002) J Am Chem Soc 124:4561
50. Rigamonti L, Demartin F, Forni A, Righetto S, Pasini A (2006) Inorg Chem 45:10976
51. Fukui H, Kishi R, Minami T, Nagai H, Takahashi H, Kubo T, Kamada K, Ohta K, Champagne B, Botek M, Masayoshi Nakano M (2008) J Phys Chem A 112:8423
52. Muller TJJ, Netz A, Ansorge M, Schmalzlin E, Brauchle C, Meerholz K (1999) Organometallic 18:5066
53. Roberto D, Ugo R, Bruni S, Cariati E, Cariati F, Fantucci PC, Invernizzi I, Quici S, Ledoux I, Zyss J (2000) Organometallic 19:1775
54. Maury O, Viau L, Senechal K, Corre B, Guegan JP, Renouard T, Ledoux I, Zyss J, Le Bozec H (2004) Chem Eur J 10:4454
55. Labat L, Lamere J-F, Sasaki I, Lacroix PG, Vendier L, Asselberghs I, Perez-Moreno J, Clays K (2006) Eur J Inorg Chem 3105

56. Nguyen P, Lesley G, Marder TB, Ledoux I, Zyss J (1997) Chem Mater 9:406
57. McDonagh AM, Humphrey MG, Samoc M, Luther-Davies B, Houbrechts S, Wada T, Sasabe H, Persoons A (1999) J Am Chem Soc 121:1405
58. Weyland T, Ledoux I, Brasselet S, Zyss J, Lapinte C (2000) Organometallic 19:5235
59. Pizzotti M, Ugo R, Dragonetti C, Annoni E (2003) Organometallic 22:4001
60. Wu K, Li J, Lin C (2004) Chem Phys Lett 388:353
61. Wu K, Sa R, Lin C (2005) New J Chem 29:362
62. Li Q, Sa R, Wei Y, Wu K (2008) J Phys Chem A 112:4965
63. Isaenko L, Yelisseyev A, Lobanov S, Petrov V, Rotermund F, Slekys G, Zondy JJ (2002) J Appl Phys 91:9475
64. Gradinaru J, Forni A, Druta V, Tessore F, Zecchin S, Quici S, Garbalau N (2007) Inorg Chem 46:884
65. Coe BJ, Jones LA, Harris JA, Brunschwig BS, Asselbergh I, Clays K, Persoons A, Garin J, Orduna J (2004) J Am Chem Soc 126:3880
66. Inerbaev TM, Belosludov RV, Mizuseki H, Takahashi M, Kawazoe Y (2006) J Chem Theory Comput 2:1325
67. Senechal-David K, Hemeryck A, Tancrez N, Toupet L, Williams JAG, Ledoux I, Zyss J, Boucekkine A, Guegan JP, Le Bozec H, Maury O (2006) J Am Chem Soc 128:12243
68. Champagne B, Perpete EA, van Gisbergen SJA, Baerends EJ, Snijders JG, Soubra-Ghaoui C, Robins KA, Kirtman B (1998) J Chem Phys 109:10489
69. van Gisbergen SJA, Schipper PRT, Gritsenko OV, Baerends EJ, Snijders JG, Champagne B, Kirtman B (1999) Phys Rev Lett 83:694
70. Jacquemin D, Andre JM, Perpete EA, Sekino H, Yasuyuki M, Kamiya M, Hirao K (2004) J Chem Phys 121:4389
71. Jacquemin D, Perpete EA, Ciofini I, Adamo C (2005) Chem Phys Lett 405:376
72. Champagne B, Perpete EA, Jacquemin D, van Gisbergen SJA, Baerends EJ, Soubra-Ghaoui C, Robins KA, Kirtman B (2000) J Phys Chem A 104:4755
73. Zhang M, Wu KC, Liu C, Wei Y (2005) Acta Phys Sinica 54:1762
74. Bruschi M, Fantucci P, Pizzotti M (2005) J Phys Chem A 109:9637
75. van Faassen M, de Boeij PL, van Leeuwen R, Berger JA, Snijders JG (2002) Phys Rev Lett 88:186401
76. Gonze X, Ghosez P, Godby RW (1995) Phys Rev Lett 74:4035
77. Champagne B, Bulat FA, Yang WT, Bonness S, Kirtman B (2006) J Chem Phys 125:194114
78. Bulat FA, Toro-Labbe A, Champagne B, Kirtman B, Yang WT (2005) J Chem Phys 123: 014309
79. Sexton JZ, Kummel AC (2003) J Vacuum Sci Technol B 21:1908
80. Lee C, Yang W, Parr RG (1988) Phys Rev B 37:785
81. Becke AD (1993) J Chem Phys 98:5648
82. Adamo C, Barone V (1998) J Chem Phys 108:664
83. Perdew JP, Wang Y (1992) Phys Rev B 45:13244
84. Hieringer W, Baerends EJ (2006) J Phys Chem A 110:1014
85. Cotton FA, Extine MW (1978) J Am Chem Soc 100:3788
86. Cai S, Hoffman DM, Wierda DA (1991) Inorg Chem 30:827
87. Becke AD (1988) Phys Rev A 38:3098
88. Perdew JP (1988) Phys Rev B 33:8822
89. Ceperly DM, Alder BJ (1980) Phys Rev Lett 45:566
90. Vosko SH, Wilk L, Nusair M (1980) Can J Phys 58:1200
91. Stoll H, Pavlidou CME, Preuss H (1978) Theor Chim Acta 49:143
92. van Lenthe E, Baerends EJ, Snijders JG (1993) J Chem Phys 99:4597
93. van Lenthe E, Baerends EJ, Snijders JG (1996) J Chem Phys 105:6505
94. van Lenthe E, Ehlers AE, Baerends EJ (1999) J Chem Phys 110:8943
95. van Gisbergen SJA, Snijders JG, Baerends EJ (1998) J Chem Phys 109:10644

96. te Velde G, Bickelhaupt FM, Baerends EJ, Fonseca-Guerra C, van Gisbergen SJA, Snijders JG, Ziegler T (2001) J Comput Chem 22:931
97. Mayer I (1983) Chem Phys Lett 97:270
98. Barclay T, Eglin JL, Smith LT (2001) Polyhedron 20:767
99. Dequeant MQ, Bradley PM, Xu GL, Lutterman DA, Turro C, Ren T (2004) Inorg Chem 43:7887
100. Golichenko AA, Shtemenko AV (2006) Russ J Coord Chem 32:242
101. Landis CR, Weinhold F (2006) J Am Chem Soc 128:7335
102. Acott SR, Garner CD, Nicholson JR, Clegg W (1983) J Chem Soc Dalton Trans 713
103. Zhu NY, Zheng YF, Wu XT (1990) Inorg Chem 29:2705
104. Klamt A (1995) J Phys Chem 99:2224
105. Klamt A, Jones V (1996) J Chem Phys 105:9972
106. Pye CC, Ziegler T (1999) Theor Chem Acc 101:396
107. Klamt A, Schüürmann G (1993) J Chem Soc Perkin Trans 2 799
108. Clays K, Persoons A (1991) Phys Rev Lett 66:2980
109. Hendrickx E, Clays K, Persoons A (1998) Acc Chem Res 31:675, and the references therein
110. Oudar JL, Chemla DS (1977) J Chem Phys 66:2664
111. Cyvin SJ, Rauch JE, Decius JC (1965) J Chem Phys 43:4083
112. Bersohn R, Pao YH, Frisch HL (1996) J Chem Phys 45:3184
113. Pizzotti M, Ugo R, Roberto D, Bruni S (2002) Organometallic 21:5830
114. Zhang C, Song YL, Kuhn FE, Wang Y, Fun H, Xin XQ (2002) J Mater Chem 12:239
115. Zhang C, Song YL, Kuhn FE, Wang Y, Xin XQ, Herrmann WA (2002) Adv Mater 14:818
116. Zhang QF, Zhan Y, Ding JH, Song YL, Rothenberger A, Fenke D, Leung WH (2006) Inorg Chem 45:5148
117. Muller A, Bogge H, Schimanski U (1983) Inorg Chim Acta 69:5
118. Chemla DS, Oudar JL, Jerphagnon J (1975) Phys Rev B 12:4534

Struct Bond (2012) 145: 81–120
DOI: 10.1007/430_2011_75

Published online: 3 March 2012

The Recent Development of SRS and SRS SF-Conversion Laser Crystal

Chaoyang Tu

Abstract Although tungstates posses lower coefficient of thermal conductivity, tungstates doped with active ions have higher quanta efficiency of fluorescence owing to their higher doping concentration of active ions resulted from the higher covalence of WO_4^{2-} units. Therefore, they are favorable for the medium of low power laser when doped with active ions. On the other hand, they have higher stimulated Raman scattering (SRS) plus in a general way. Therefore, tungstates doped with active ions can serve as a SRS self-frequency conversion multifunctions laser medium. After wide surveys of known research on the growth, crystal structure, and properties including optical and spectra characteristics and laser property, this chapter reviews the recent advances in the development of KGd $(WO_4)_2$ and $SrWO_4$ Raman and SRS self-frequency conversion laser crystal. The SRS self-frequency conversion laser technology was dealt with. As a result, the Raman and self-Raman laser outputs with high efficiency at ~1,180 nm wavelength and its frequency-doubling laser outputs at ~590 nm wavelength have been achieved.

Keywords Crystal growth · $KGd(WO_4)_2$ and $SrWO_4$ crystals · Optical properties · Raman and SRS self-frequency conversion laser properties

Contents

C. Tu (✉)
Key Laboratory of Photoelectric Materials Chemistry and Physics of CAS, Fujian Institute of Research on the Structure of Matter, Chinese Academy of Sciences, Fuzhou, Fujian 350002, China
e-mail: tcy@ms.fjirsm.ac.cn

1 Introduction

Owing to their advantages including the compactness of the devices, efficiency, low maintenance and low cost, all solid-state laser sources are of special importance for a variety of applications: color projection, high density optical data storage, laser printing, medicine, photodynamic therapy, military devices, biotechnology, submarine communications, etc. Moreover, many of these laser plus materials can be pumped by commercially available laser diodes, such as around 800 or 900 nm.

Stimulated Raman scattering (SRS) has been well studied [1–6], and is applied to high-resolution nonlinear spectroscopy [7], control of coherence between quantum states [8], chirped femtosecond soliton-like laser pulse [4], and so on. Among the above applications, the development of compact solid-state Raman lasers for new important near infrared and eye-safe spectral region of 1.2–1.5 μm [9, 10] attracts a great deal of interest.

Many rare earth ions or transition ions doped in the laser host materials can serve as the emission centers to produce laser with the different specific wavelengths directly. For example, some specific wavelengths can be achieved by direct laser operation of the most well-known luminescent ions: Nd^{3+}(1.06, 1.3 μm), Yb^{3+}(1.03 μm), Er^{3+}(1.5 μm), Tm^{3+}(1.8 ~ 2.0 μm), Ho^{3+}(2.0 μm), Cr^{3+}(0.7 ~ 1.0 μm) etc. incorporated into a suitable laser plus crystal or glass host. Furthermore, more new specific wavelengths can be obtained from a third-order nonlinear optical interaction in Raman laser plus crystals, such as $KGd(WO_4)_2$ (KGW), $BaWO_4$, $SrWO_4$, $La_2(WO_4)_3$ [11], YVO_4, $GdVO_4$ [12], $Ba(NO_3)_2$, $PbWO_4$ [13].

If the laser effect and the third-order nonlinear Raman optical phenomena occur simultaneously inside the same crystal, simplifying the devices, a simplified and attractive scheme namely the Stimulated Raman scattering (SRS) self-frequency conversion process is obtained. Only if the crystal can provide the localities for luminescent ions, such as $KGd(WO_4)_2$ (KGW), $SrWO_4$ crystals , an optical nonlinear crystal can serve as the SRS self-frequency conversion crystal.

In this chapter, the study on the Raman and SRS self-frequency conversion laser crystals of pure and rare earth ions doped $KGd(WO_4)_2$ and $SrWO_4$ crystals is covered. The growth, crystal structure, optical, and spectrum characteristics of these crystals are presented. The Raman and SRS self-frequency conversion laser characteristics are also presented.

2 Self-Frequency Conversion Laser Crystal Re^{3+}:$KGd(WO_4)_2$

The $KGd(WO_4)_2$ (KGT) crystal has a structure belonging to the monoclinic system with space group C2/c and unit cell dimensions $a = 80$ Å, $b = 10.43$ Å, $c = 7.6$ Å [14]. This crystal is not only an excellent host material for solid state lasers but also a Raman laser material. The Raman gain coefficient of KGT at $\lambda = 1{,}064$ nm is 4.1 (cm/GW). The threshold of the laser oscillations in an Nd^{3+}-doped KGT laser crystal is considerably low [15]. The fluorescence concentration quench effect of the Nd^{3+} ion in the KGT crystal may be weakened because of higher covalent characteristics of the W–O covalent bond. Therefore, a higher Nd^{3+} doping concentration can be accepted in KGT crystal. Furthermore, the absorption band at 0.8 μm of Nd^{3+} ion in the KGT crystal is well matched with the emission wavelength of a laser diode, a solid-state laser pump which is very convenient and popular in laser science and technology today. Rare earth ions-doped KGT laser crystal also can serve as a stimulated Raman scattering self-frequency conversion of radiation to a new spectral range.

2.1 *The Growth of Nd^{3+}:$KGd(WO_4)_2$ Crystal*

2.1.1 The Selection of Nd^{3+} Doping Concentration

In general, the fluorescence intensity or emission cross section will increase with the increase of Nd^{3+} doping concentration at lower Nd^{3+} doping concentration in the host, then the fluorescence quenching will engender with the further enhancement of Nd^{3+} doping concentration, which will decrease the fluorescence intensity or emission cross-section and shorten the Fluorescence lifetime. Thus, there is a finest Nd^{3+} doping concentration. Here, the measured relationship between the Fluorescence intensity and Nd^{3+} doping concentration in $KGd(WO_4)_2$ host is shown in Fig. 1, from which the suitable Nd^{3+} doping concentration is measured to be 3–8 at.% [16].

2.1.2 The Selection of Flux

Tm^{3+}, Nd^{3+}:$KGd(WO_4)_2$ crystal should be grown by flux method because $KGd(WO_4)_2$ crystal has a phase transformation at 1,005°C. K_2WO_4 and $K_2W_2O_7$ have been used as the flux. Figure 2 shows the melting temperature curve of $KGd(WO_4)_2$–K_2WO_4 system [16] and Fig. 3 shows the one of $KGd(WO_4)_2$–$K_2W_2O_7$ system [17]. The growth results demonstrated the distinctions of growth characteristics. In the $KGd(WO_4)_2$–K_2WO_4 system, K_2WO_4 flux floats on the surface of the melt whereas $KGd(WO_4)_2$ solute deposits at the bottom of the crucible because both the melting point and the density of $KGd(WO_4)_2$ are higher

Fig. 1 The measured relationship between the fluorescence intensity and Nd^{3+} doping concentration in $KGd(WO_4)_2$ host

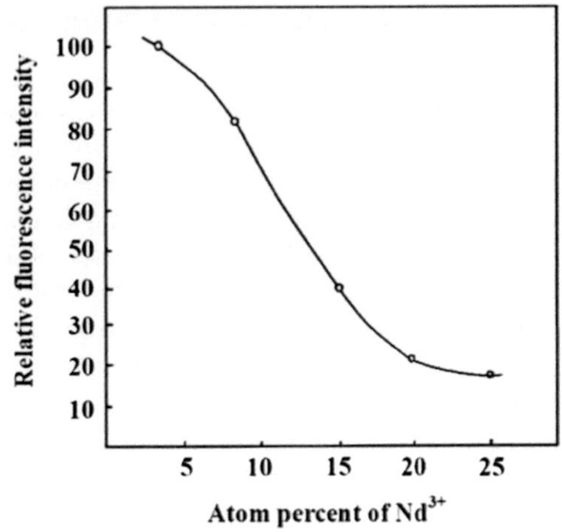

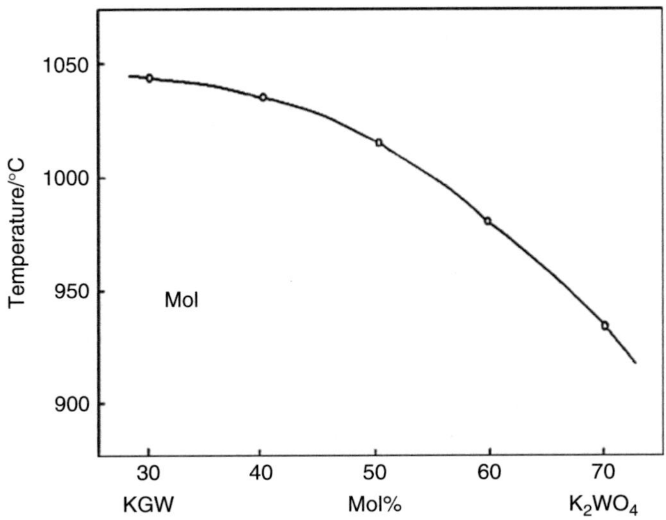

Fig. 2 The melting temperature curve of $KGd(WO_4)_2$–K_2WO_4 system

than those of K_2WO_4. Therefore, the long melt interval and the furious agitation should be introduced. Although the above phenomenon disappear in the KGd $(WO_4)_2$–$K_2W_2O_7$ system, the growth instability is aroused from the larger temperature coefficient of concentration when the content of solute is higher than 20 mol%, which results in the abundance precipitation of solute with small decrease of temperature. Therefore, the content of solute with lower than 20 mol% has to be introduced, which restrains the dimension of crystal badly.

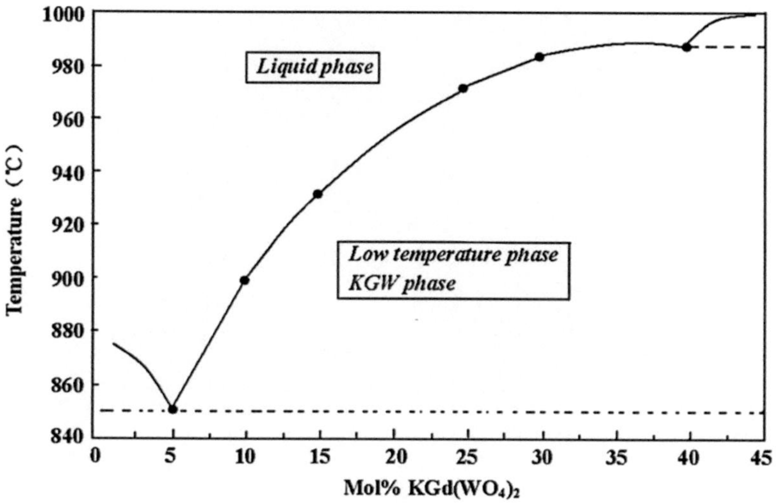

Fig. 3 The melting temperature curve of $KGd(WO_4)_2$–$K_2W_2O_7$ system

2.1.3 The Crystal Growth

The Tm^{3+}, Nd^{3+}:$KGd(WO_4)_2$ crystals were grown based on the top-seeded solution growth (TSSG) [16, 18] and TSSG combined with ACRT [19], which introduces the furious agitation, respectively. The starting materials with 60 mol% K_2WO_4 were prepared in proportions of Gd_2O_3:K_2CO_3:WO_3 = 1:1:4 mol and Tm_2O_3, (Nd_2O_3):Gd_2O_3 = 5:95 mol and they were mixed and placed into a crucible. This mixture of starting material was fired at a temperature, which was 50°C higher than the crystallization temperature and was kept at this temperature for 3–4 days to let the solution melt completely and homogeneously. The saturation temperature was exactly measured by the method of repeated seeding. A seed was placed into the melt at a temperature of 30°C above the saturation temperature for half an hour to dissolve the outer surface of the seed. When the growth process was ended, the crystal was drawn out of the melt surface and cooled down to room temperature at a rate of 50°C/h. Figures 4–6 show the as-grown crystals.

The crystal displays many faces $[\bar{1}11][010][101][11\bar{1}][110][3\bar{1}0][310][1\bar{1}0]$ $[3\bar{1}0][130]$ at the initial stages of TSSG as shown in Fig. 7 [20]. When the crystals grow up, these faces disappear and the crystals become round in the case of using the TSSG, whereas some faces remain and the crystals display square shape in the case of using modified TSSG. The structure of $KGd(WO_4)_2$ crystal consists of the WO_4 tetrahedron, distorted KO_{12} dodecahedron and GdO_8 octahedra. These polyhedrons couple each other by O^{2-} ions. The crystal faces pointing to the ridge and faces of WO_4 tetrahedron develop slowly and thus emerge as the lathing faces, whereas the crystal faces pointing to the tips of WO_4 tetrahedron develop quickly and thus disappear.

Fig. 4 The as-grown Nd^{3+}:$KGd(WO_4)_2$ crystal by modified TSSG method

Fig. 5 The as-grown Nd^{3+}:$KGd(WO_4)_2$ crystal by TSSG method

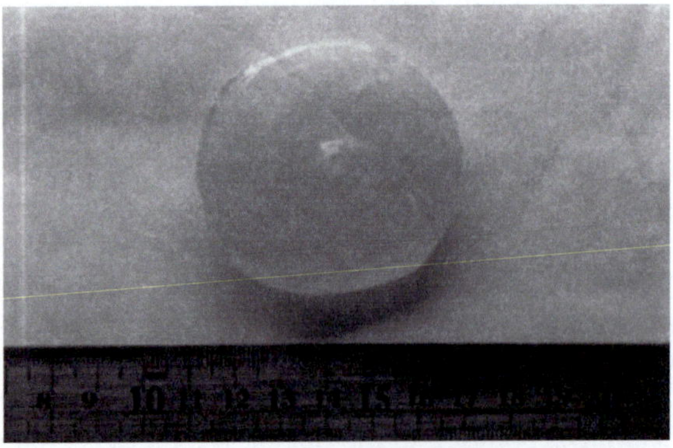

Fig. 6 The as-grown Tm^{3+}:$KGd(WO_4)_2$ crystal by TSSG method

Fig. 7 Growth morphology of the crystal

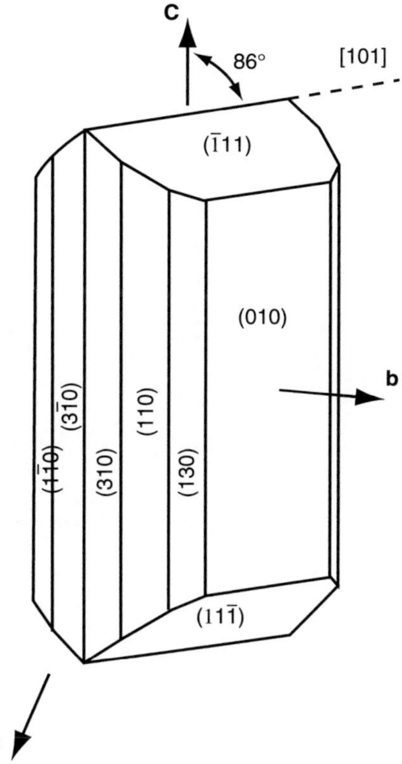

2.2 *The Spectrum Characteristics of Nd^{3+}:$KGd(WO_4)_2$ Crystal*

Figure 8 presents the Raman spectrum of Nd^{3+}:$KGd(WO_4)_2$ crystal [21], in which the absorption peaks centered at 923, 889, 835, 771, and 745 cm^{-1} are belong to flex vibration and the one centered at 400.4, 369.5, and 344.8 cm^{-1} are belong to the inflect vibration. Table 1 also presents the ascription of Raman spectrum.

Figure 9 presents the room temperature polarized absorption spectra of Nd^{3+}:$KGd(WO_4)_2$ crystal measured by Dr. A. Brenier [22], in which there are three strong absorption peaks centering at 585 nm, 752 nm, and 808 nm, respectively. The absorption cross section at 808 nm is 2.5×10^{-19} cm^2. The room temperature emission spectra of Nd^{3+}:$KGd(WO_4)_2$ crystal is presented in Figs. 10 and 11. The stimulated emission cross section at 1,340 nm corresponding to ${}^4F_{3/2} \rightarrow {}^4I_{13/2}$ transition for *Pol.Nm* and *Pol.Ng* is about $9 \times 10_{-20}$ cm^2 and 6×10^{-20} cm^2, respectively. The stimulated emission cross-section at 1064 nm corresponding to ${}^4F_{3/2} \rightarrow {}^4I_{11/2}$ transition is about 2.3×10^{-19} cm^2. The fluorescence lifetime is 119 μs [23].

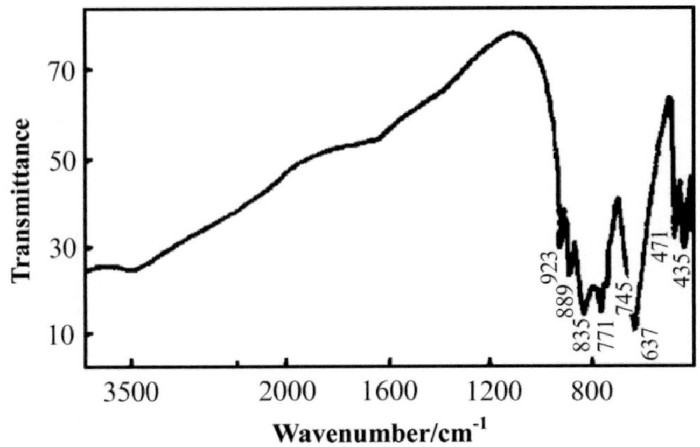

Fig. 8 The Raman spectrum of Nd^{3+}:$KGd(WO_4)_2$ crystal

Table 1 IR and Raman spectra and the relative adscription

IR (peak/nm)	Raman (peak/nm)	Adscription
	344.8	The inflect vibration of WO_4
	369.5	
	400.4	The vibration of Ln-O
408(γ2)		
428(γ2)		The flex vibration of WO_4
435	756.4	
471	766.2	
771		
	807.1	
835		
889		
	899.5	

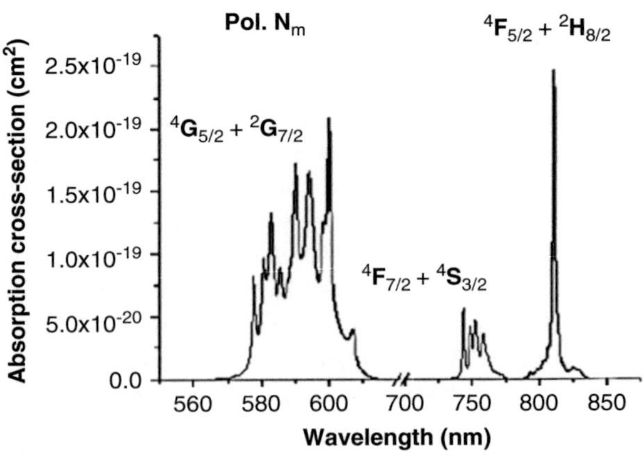

Fig. 9 The room temperature polarized absorption spectra of Nd^{3+}:$KGd(WO_4)_2$ crystal

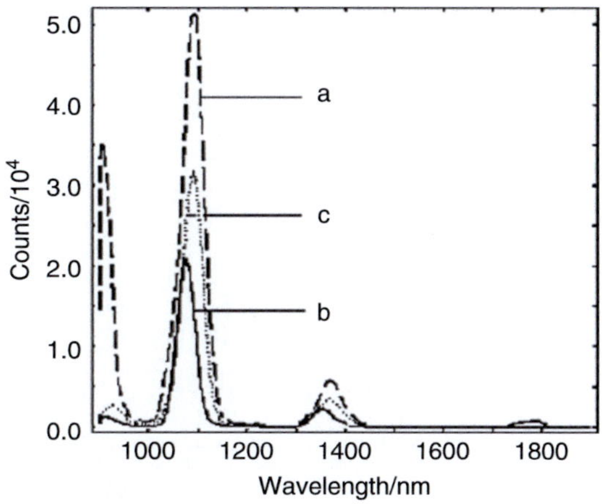

Fig. 10 The room temperature emission spectra of Nd^{3+}:$KGd(WO_4)_2$ crystal

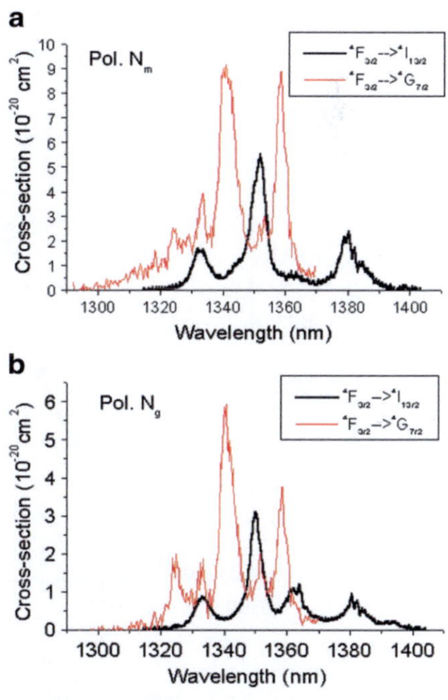

Fig. 11 The room temperature polarized emission spectra of Nd^{3+}: $KGd(WO_4)_2$ crystal

2.3 The Spectrum Characteristics of Tm^{3+}:$KGd(WO_4)_2$ Crystal

Figure 12 presents the room temperature absorption spectra of Tm^{3+}:$KGd(WO_4)_2$ crystal [24]. The room temperature emission spectra of Tm^{3+}:$KGd(WO_4)_2$ crystal is presented in Fig. 13. The absorption cross section at 800 nm is 3.2×10^{-20} cm^2. The stimulated emission cross section at 1,800 nm is about 1.56×10^{-20} cm^2. Table 2 gives the J–O strength parameters of some Tm-doped crystals. Table 3 presents

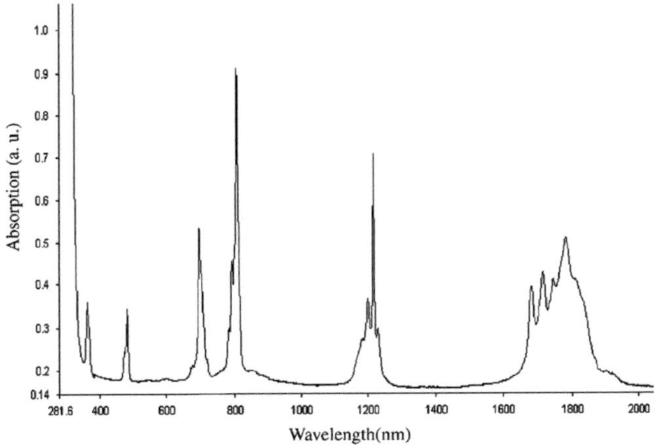

Fig. 12 The room temperature absorption spectra of Tm^{3+}:$KGd(WO_4)_2$ crystal

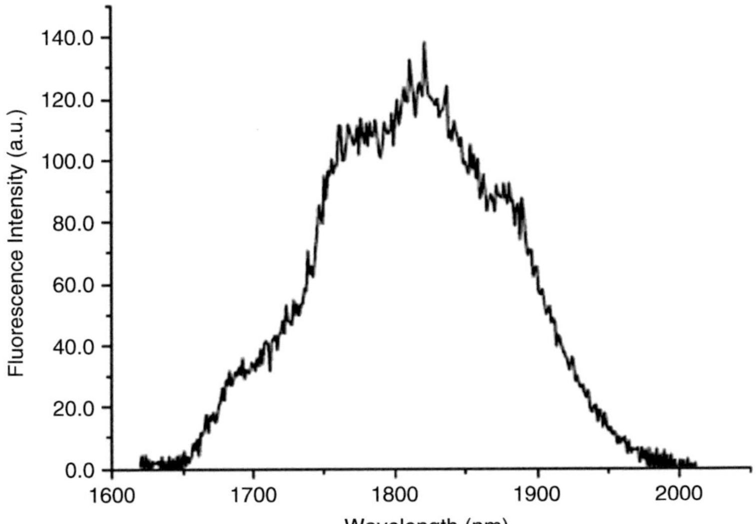

Fig. 13 The room temperature emission spectra of Tm^{3+}:$KGd(WO_4)_2$ crystal

Table 2 The Judd–Ofelt strength parameters for Tm^{3+}-doped crystals

Crystal	$\Omega_2 \times 10^{20}$	$\Omega_2 \times 10^{20}$	$\Omega_2 \times 10^{20}$	Ref.
Tm^{3+}:$KGd(WO_4)_2$	2.64	5.84	14	This work
Tm^{3+}:YVO_4	1.94	0.158	0.396	[20]
Tm^{3+}:$SrGd(Ga_3)O_7$	1.29	1.08	0.47	[21]
Tm^{3+}:$CaYAlO_4$	1.55	3.45	1.18	[22]
Tm^{3+}:Y_2O_3	4.07	1.46	0.61	[23]
Tm^{3+}:YVO_4	9	1.05	2.27	[23]

Table 3 The calculated radiative transition rates, the fluorescence branching ratios, and the radiative lifetimes for different transition levels

Start levels	Terminating levels	Transition wavelength (nm)	A (s^{-1})	$\beta(J') = \frac{A(J \to J')}{A_T(J)}$ (%)	τ_R
$^1G_4$	$^3F_2$	1,678	46.3	0.2	33 μs
	$^3F_3$	1,531	870.4	2.9	
	$^3F_4$	1,182	426.7	1.43	
	$^3H_5$	779	10,923.8	36.6	
	$^3H_4$	665	13,352.1	44.7	
	$^3H_6$	481	4,323.2	14.5	
$^3F_2$	$^3F_3$	31,056	0.018	0	48 μs
	$^3F_4$	5,230	14.8	0.1	
	$^3H_5$	1,559	3,264.2	15.8	
	$^3H_4$	1,109	2,690.2	13.0	
	$^3H_6$	675	14,687.6	71.1	
$^3F_3$	$^3F_4$	6,289	8.3	0	24 μs
	$^3H_5$	1,447	1,081.2	2.6	
	$^3H_4$	1,150	3,783.4	9.0	
	$^3H_6$	689	37,038.4	88.4	
$^3F_4$	$^3H_5$	2,222	101.7	0.7	66 μs
	$^3H_4$	1,408	1,055.4	7.0	
	$^3H_6$	775	14,025.5	92.4	
$^3H_5$	$^3H_4$	3,842	29.8	0.8	0.3 ms
	$^3H_6$	1,190	3,550.1	99.2	
$^3H_4$	$^3H_6$	1,724	1,240.8	100	0.8 ms

calculated radiative transition rates, the fluorescent branching ratios, and radiative lifetimes of Tm^{3+} ion. Table 4 presents the theoretical parameters for the transitions between the energy levels of Tm^{3+} ion. Table 5 shows the experimental oscillator strengths $P_{iq}^{(exp)}$and theoretical oscillator strengths $P_{iq}^{(cal)}$.

2.4 *The Stimulated Raman Scattering Frequency Self-Conversion*

A short-pulsed longitudinal pumping at 600 nm has been used and SRS frequency self-conversion of the 1,352.5 nm laser line in $KGd(WO_4)_2$:Nd^{3+} was obtained by Dr. Alain Brenier [22]. The wavelength, which was produced is eyesafe at 1,539.5 nm, and the conversion efficiency was 1.26% with uncoated faces of the crystal. The yield can be improved by using antireflection-coated faces, an input mirror highly reflective at the Stokes 1 wavelength and an output mirror with a larger radius of curvature more adapted to our pump waist. Although nonoptimal experimental conditions, a higher conversion efficiency than that in previous SRS self-conversion in the eye-safe region (0.15%) was reached. Figure 14 presents the time evolution for the pump at 600 nm, laser at 1,068.2 nm, and Stokes 1 SRS pulse at 1,181 nm. Figure 15 presents the energies of the laser at 1,068 nm and the Stokes 1

Table 4 The experimental oscillator strength $P_{iq}^{(exp)}$ and theoretical oscillator strengths $P_{iq}^{(cal)}$ for Tm^{3+} transition from the $^3H_6$ ground state

Wavelength (nm)	$P_{iq}^{(exp)} \times 10^5$	$P_{iq}^{(cal)} \times 10^5$
1,742	1.38	1.06
1,215	0.96	1.49
795	2.23	1.55
689.4	2.33	2.42
475.6	2.38	2.38
358.64	3.38	3.14

Table 5 Theoretical parameters for the transitions between the energy levels of Tm^{3+} ions

Transition levels	Center wavelength (nm)	Radiative transition rates (s^{-1})	$\Delta\nu$ (cm^{-1})	Stimulated emission cross-section, σ_p (cm^2)
$^1G_4 \rightarrow {}^3H_6$	455	4,323.2	298.5	6.08×10^{-21}
$^1G_4 \rightarrow {}^3H_4$	665	13,352.14	500	2.40×10^{-20}
$^1H_4 \rightarrow {}^3H_6$	1,825	1,240.77	538.8	1.56×10^{-20}

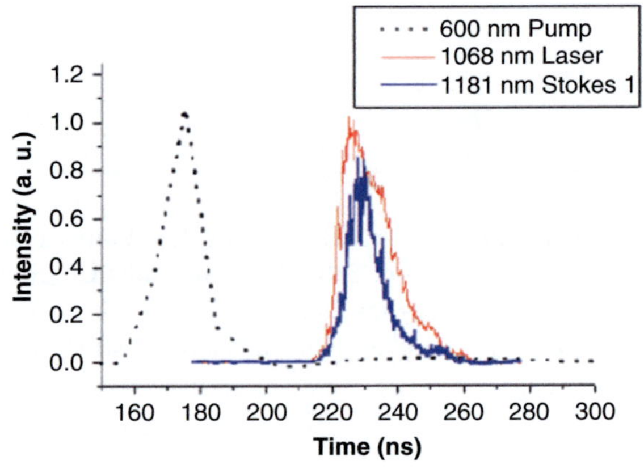

Fig. 14 The time evolution for the pump at 600 nm, laser at 1,068.2 nm and Stokes 1 SRS pulse at 1,181 nm

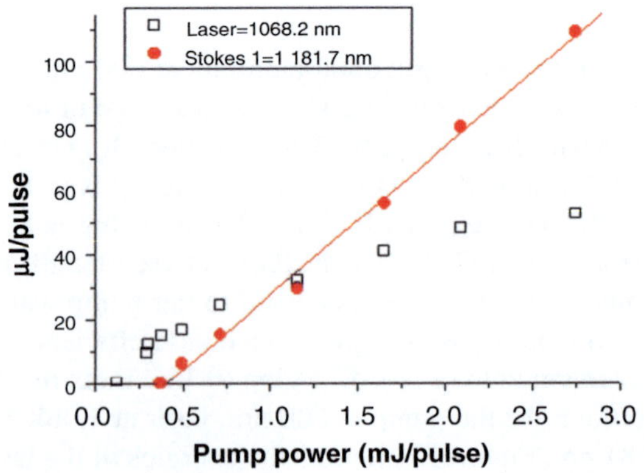

Fig. 15 Energies of the laser at 1,068 nm and the Stokes 1 SRS pulses vs. the pump power

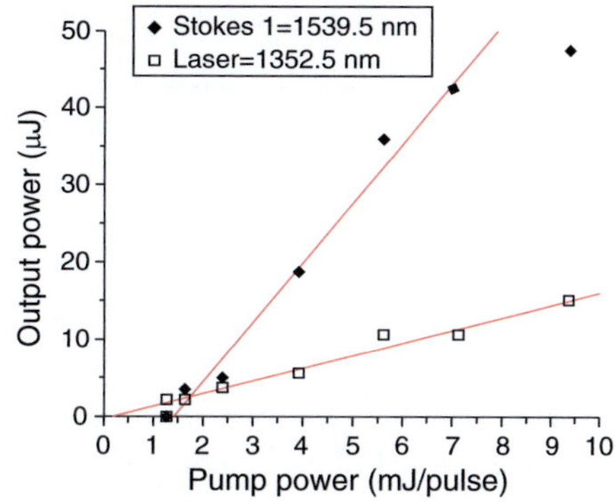

Fig. 16 The energies of the laser at 1,352.5 nm and the Stokes 1 SRS at pulses 1,539.5 nm vs. the pump power

SRS pulses versus the pump power. Figure 16 shows the energies of the laser at 1,352.5 nm and the Stokes 1 SRS at pulses 1,539.5 nm versus the pump power.

A solid self-Raman laser was reached by Prof. Lin based on a Nd:KGW crystal that is transversely pumped by laser diode bars [25]. A beam of an eye-safe laser with a 31.8 mJ output energy and a 2.0 ns pulse width was obtained by applying a special *s*-polarized reflective resonator configuration in which the length of the Raman resonator was shorter than that of the fundamental radiation resonator. The eye-safe laser has the highest output energy and the shortest pulse width among the Nd:KGW lasers ever reported.

3 SRS Self-Frequency Conversion Laser Crystal $Nd^{3+}:SrWO_4$

$SrWO_4$ belongs to the scheelite family, the unit cell parameters are: $a = 5.4168$ Å, $c = 11.951$ Å, $V = 350.66$ Å^3, $z = 4$, $Dc = 6.35$ g/cm^3 and the space group is 85I4$_1$/a [26]. As a new promising Raman-active crystal, $SrWO_4$ has attracted much attention for its good mechanical and optical properties. Its steady-state Raman gain coefficient is 5.0 cm/GW at 1,064 nm. The Raman mode at 921 cm^{-1} has the strongest intensity, and its line width is 2.7 cm^{-1} at room temperature [27]. Table 6 shows the Raman properties of the some Raman materials. The spontaneous Raman spectroscopy of $SrWO_4$ as the Raman-active material was studied using an alexandrite free-running multimode laser as the pump source [27]. Furthermore, the crystal structure of tungstate crystals allows this crystal as matrices for laser-active elements with nonlinear self-conversion of radiation to a new spectral range [27].

3.1 The Crystal Growth

It is well known that $SrWO_4$ crystal belongs to the uniaxial, tetragonal structure with a space group of $I4_1/a$ (C_{4h}^6). The scheelite unit cells contain four molecules of

Table 6 Raman properties of the materials

Property	$SrWO_4$	$BaWO_4$	$KGd(WO_4)_2$	$Ba(NO_3)_2$
Raman frequency (cm^{-1})	921	926	901	1,047
Raman line width (cm^{-1})	2.7	1.6	5.4	0.4
Integral cross-section (%)	–	47	54	21
Peak scattering cross-section (%)	30	64	25	63
Threshold (mJ)	6.5	4.2	7.4	2.8
Raman gain coefficient, $\lambda = 1{,}064$ nm (cm/GW)	5.0	8.5	4.1	11

$SrWO_4$ and the strontium ions coordinate with eight oxygens, each of which is a member of a $(WO_4)^{2-}$ complex [26]. Measurements of electron paramagnetic resonance carried out in [28] indicate that the rare-earth ions are located in an undistorted cation site of which point symmetry is S_4. As in this tungstate the trivalent rare-earth ion is introduced on a divalent cation site, a mechanism for charge compensation should be presented to hold roughly for the scheelite. Therefore, Nb_2O_5 was added to the melt in the ratio of one neodymium to one niobium to improve its optical properties [27].

Nd^{3+}-doped $SrWO_4$ crystals were grown in N_2 atmosphere along <001>direction by using Czochralski method [29]. The chemicals used were analytical grade $SrCO_3$, WO_3, and spectral grade Nd_2O_3. The starting materials were prepared by mixing $SrCO_3$, WO_3, and Nd $_2O_3$ powders according to reaction formula:

$$SrCO_3 + \ x\ Nd_2O_3 + \ WO_3 \rightarrow 2\ Nd : SrWO_4 + CO_2 \uparrow$$

The weighed materials with doping 1 ~ 2at.% Re^{3+} were thoroughly mixed and pressed and put into a platinum crucible with $\Phi 50 \times 50\ mm^3$, then heated to 750°C and kept for 18 h to decompose the $SrCO_3$, and ground, mixed again, and then reheated to 1,200°C, kept for 24 h. The obtained sample was very hard ceramics.

The synthesized material melted congruently at 1,450°C. The iridium crucible was heat by conventional RF-heating method. Crucible size is 50 mm in diameter and 50 mm in height. The pulling rate was 1–1.5 mm/h and the crystal rotation rate at 12–20 rpm. To release the stress produced in the temperature-lowering process, the crystals were annealed at 1,200°C for 5–6 h and then cooled down to room temperature at a rate of 20 K/h. In order to avoid the cracking of the crystal, the above crystal must be taken to anneal again in O_2 atmosphere according to a special temperature-controlled procedure.

As a result, high-quality Nd doped-$SrWO_4$ crystals with dimension of $\Phi 20$ mm $\times$ 70 mm (shown in Fig. 17) was obtained. The result shows that its optical homogeneity is 4×10^{-5}. It shows that it has excellent quality. Table 7 presents the growth parameters of $SrWO_4$ crystals.

In order to obtain the large-sized Nd doped-$SrWO_4$ crystals with high optical homogeneity, the control of growing processes and conditions are very important first. Then the used raw materials must be highly pure. Second, to get the defined composition of the melt, the preparation of chemicals was found to be important.

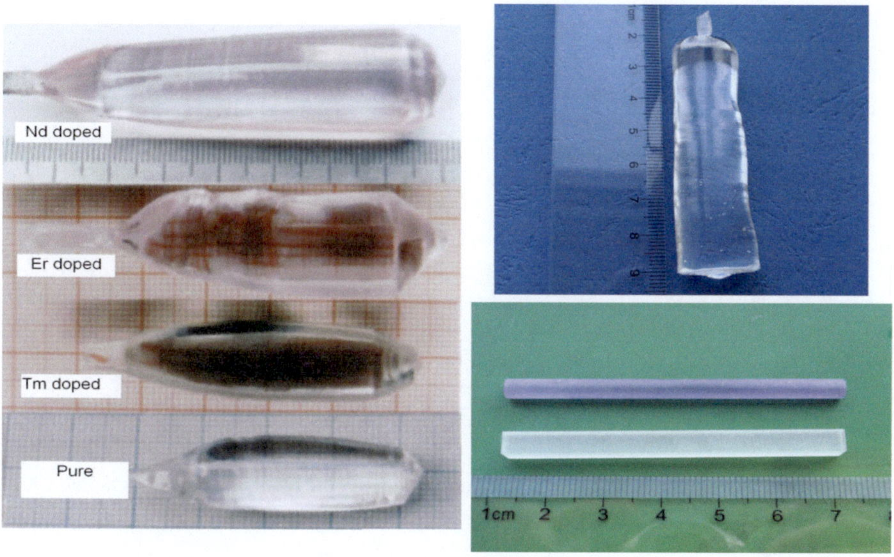

Fig. 17 As-grown pure $SrWO_4$ and rare earth doped-$SrWO_4$ crystals

Table 7 Growth parameters of Nd :$SrWO_4$ crystals

Parameters	Property	Parameters	Property
Seeding temp. (°C)	1,540	Atmosphere	N_2
Soaping temp. (°C)	1,590	Pulling rate (mm/h)	1–1.2
Soaping time (h)	2	Rotate rate (rpm)	12–15
Crucible size (mm)	$\phi = 50 \times 50$	Annealing rate (°C/h)	12–30
Crucible	Iridium		

Third, the seed surface must be melted to remove the defect in the seed before growing. And the growth point temperature must be a little higher than the saturation point. The control of the pulling rate, rotating rate, and annealing rate is also very important. Finally, a designed after-heater should be used to avoid the crack of large crystal.

3.2 *The Thermal Characteristic*

Measurements of thermal expansion have greatly increased our knowledge of material properties such as lattice dynamics, electronic and magnetic interactions, thermal defects, and phase transitions [29]. As a significant part of the power pump is converted into heat inside the material during laser operation, it is important to know its linear thermal expansion coefficients to predict how the material behaves

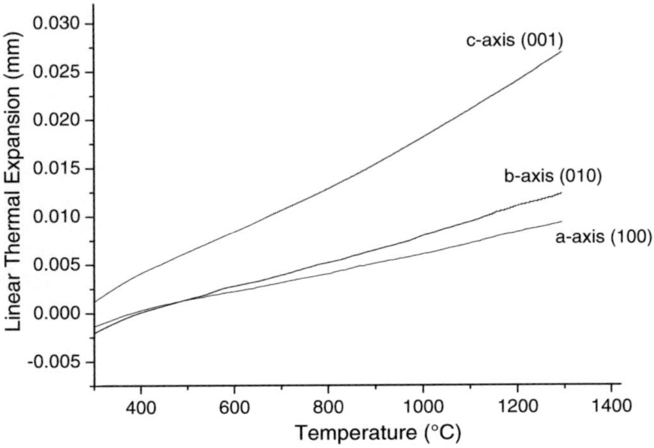

Fig. 18 Linear thermal expansion in a function of temperature along the crystallographic (001), (100), (010) directions

when the temperature increases [30]. The sample lengths of the three crystallographic *a*-, *b*-, and *c*-axes were 5.78 mm, 3.78 mm, and 4.86 mm, respectively. The figure of linear expansions versus temperature was shown in Fig. 18. The linear thermal expansion coefficient is defined as:

$$\alpha = \frac{1}{T}\frac{\Delta L}{\Delta T}, \tag{1}$$

where L is the initial length of the sample at room temperature and ΔL is the change in length when the temperature changes ΔT. We can calculate the thermal expansion coefficient from the slope of the linear fitting of the linear relationship between $\Delta T/T$ and the temperature. In this case, the linear thermal expansion coefficients for different crystallographic direction *c*-, *b*- and *a*-axes are 2.73×10^{-5}°C^{-1}, 1.00×10^{-5}°C^{-1}, and 1.05×10^{-5}°C^{-1}, respectively. The thermal expansion coefficients obtained in this work are comparable to those of $BaWO_4$ crystal reported in [31]. The comparison of thermal expansion coefficients between $SrWO_4$ and $BaWO_4$ crystals is shown in Table 8.

In a uniaxial scheelite structure $SrWO_4$ crystal, the values of thermal expansion coefficients perpendicular to the crystallographic axes (*c*-axis) are theoretical equivalent. The value of thermal expansion along *c*-axis is about two times larger than that of *a*- and *b*-axis, and the value along *a*- and *b*-axes are comparable. The small difference of thermal expansion between *a*- and *b*-axes was attributed to experimental errors.

$$\alpha_{ij(\mathrm{SrWO_4})} = \begin{vmatrix} 1.05 & 0 & 0 \\ 0 & 1.00 & 0 \\ 0 & 0 & 2.73 \end{vmatrix} \times 10^{-6\circ}\mathrm{C}.$$

Table 8 The comparison of thermal expansion coefficients between $SrWO_4$ and $BaWO_4$ crystals

Crystals	Average thermal expansion coefficient (°C^{-1})			Ref.
	(001)	(010)	(100)	
$SrWO_4$	2.73×10^{-5}	1.00×10^{-5}	1.05×10^{-5}	[32]
$BaWO_4$	3.12×10^{-5}	0.89×10^{-5}	0.82×10^{-5}	[31]

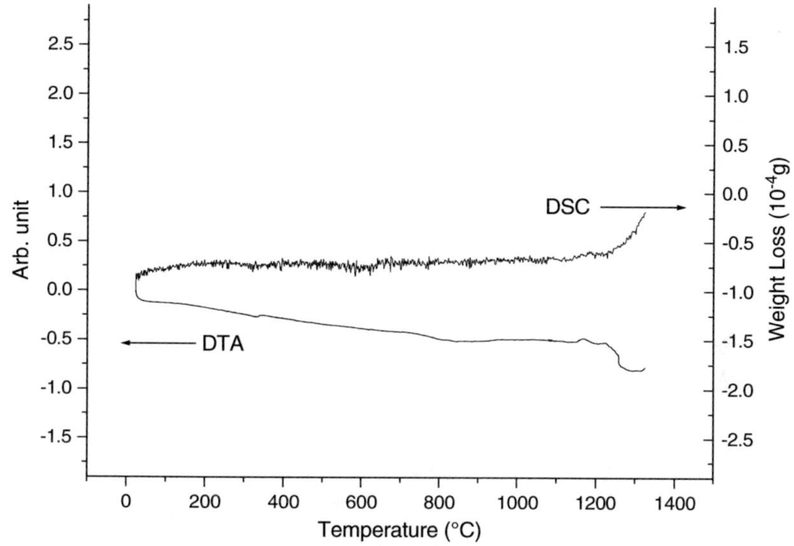

Fig. 19 DTA and DSC graphs of pure $SrWO_4$ crystal

Figure 19 shows the DTA and DSC graphs of pure $SrWO_4$ crystal. A decalescence peak in DTA curve corresponding to the agravic peak in DSC curve may be resulted from the volatilization of superfluous WO_3, which is similar to the phenomena in $La_2(WO_4)_3$ and $Sm_2(WO_4)_3$ [11, 33] .

3.3 The Optical and Spectroscopic Characteristics

3.3.1 Measurements of Refractive Index

The minimum deviation technique with a prism sample is a popular method used in the refractive index measurements [34]. To fully characterize Nd^{3+}:$SrWO_4$ in terms of refractive indices as a function of wavelength, we have measured it with a precision of 1×10^{-4} at several different wavelengths from the visible to the near infrared. These data were fitted by the least-square method according to the Sellmeir equation [35]:

Table 9 Measured and calculated refractive indices of $SrWO_4$ crystal

λ (μm)	n_o	n_e
0.8080	1.87649	1.8678
0.6563	1.88701	1.87762
0.6328	1.88966	1.88072
0.5893	1.89472	1.88525
0.5780	1.89610	1.88669
0.5461	1.90077	1.89113
0.5321	1.90328	1.89298
0.4861	1.91294	1.90228
0.4358	1.92666	1.91574

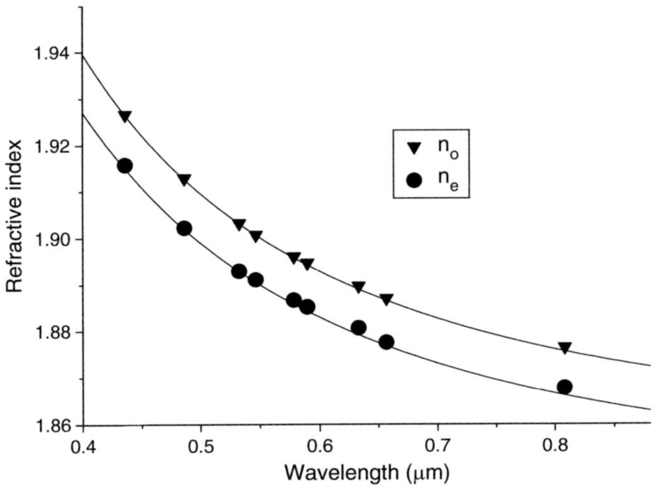

Fig. 20 Dispersion of the refractive indices and calculated curves from the Sellmeir coefficients for $SrWO_4$

$$n_i^2 = A + \frac{b}{\lambda^2 - C} - D\lambda^2, \tag{2}$$

where λ is the wavelength, and A, b, C, and D are the Sellmeir parameters. By fitting the datas reported in Table 9, the following equations (λ is in the unit of μm) are obtained:

$$n_o^2 = 3.483 + \frac{0.0523}{\lambda^2 + 0.001523} + 0.0049856\lambda^2 \tag{3}$$

$$n_e^2 = 3.4033 + \frac{0.0526}{\lambda^2 + 0.009456} + 0.0096059\lambda^2. \tag{4}$$

Figure 20 shows the dispersion of the refractive indices and calculated curves from the Sellmeir coefficients for $SrWO_4$. The differences between the measured and the calculated values are less than 10^{-4}. It is the first time that the refractive indices of $SrWO_4$ crystal are precisely measured. The values of refractive index are employed in the spectroscopic calculations.

3.3.2 The Spectrum Characteristics of Nd^{3+}:$SrWO_4$ Crystal

The room temperature polarized absorption spectra of Nd^{3+}:$SrWO_4$ crystal are presented in Fig. 21 [36], in which there are three strong absorption peaks centering at 585 nm, 752 nm, and 805 nm, respectively. The FWHMs of the absorption bands for both σ and π polarizations at 805 are 8 nm. The room temperature polarized emission spectra of Nd^{3+}:$SrWO_4$ crystal are presented in Fig. 22, in which there are three strong emission bands centering at 890 nm, 1,060 nm, and 1,344 nm corresponding to ${}^4F_{3/2} \rightarrow {}^4I_{9/2}$, ${}^4F_{3/2} \rightarrow {}^4I_{11/2}$, and ${}^4F_{3/2} \rightarrow {}^4I_{13/2}$ transitions, respectively. The cross sections at 1,060 nm for σ and π polarizations are 1.75×10^{-20} cm^2 and 2.04×10^{-20} cm^2, respectively. Table 10 presents the comparison of the spectroscopic parameters of Nd^{3+}:$SrWO_4$ crystal and other Nd^{3+}- doped tungstate crystals. Table 11 shows the calculated radiative transition rate, the branching ratios, and the radiative lifetime for different transition levels. Figure 23 presents the room temperature fluorescence decay curve of Nd^{3+}:$SrWO_4$ crystal. The lifetime of ${}^4F_{3/2}$ level is about 188.80 μs and relative luminescent quantum efficiency is about 95.23%. Compared with Nd:$^{3+}$:$PbWO_4$ 165/175 μs [41], 1% Nd^{3+}:$KY(WO_4)_2$ 125 μs, 5% Nd^{3+} :$KY(WO_4)_2$ 70 μs [37], 1% Nd $^{3+}$:KGd $(WO_4)_2$ 120 μs [38], Nd^{3+}:$SrWO_4$ crystal has the highest lifetime.

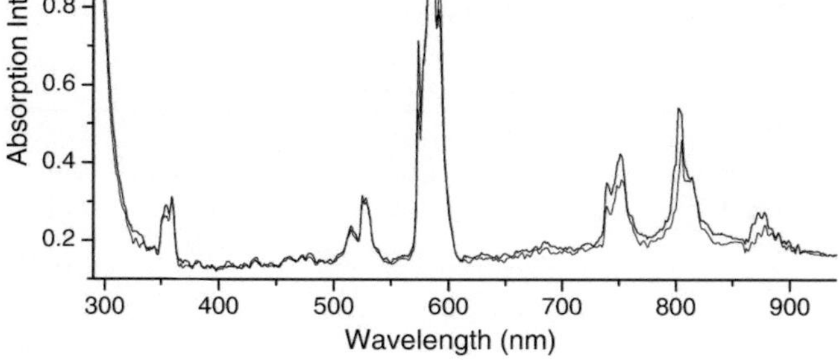

Fig. 21 Room temperature-polarized absorption spectra of Nd^{3+}:$SrWO_4$ crystal

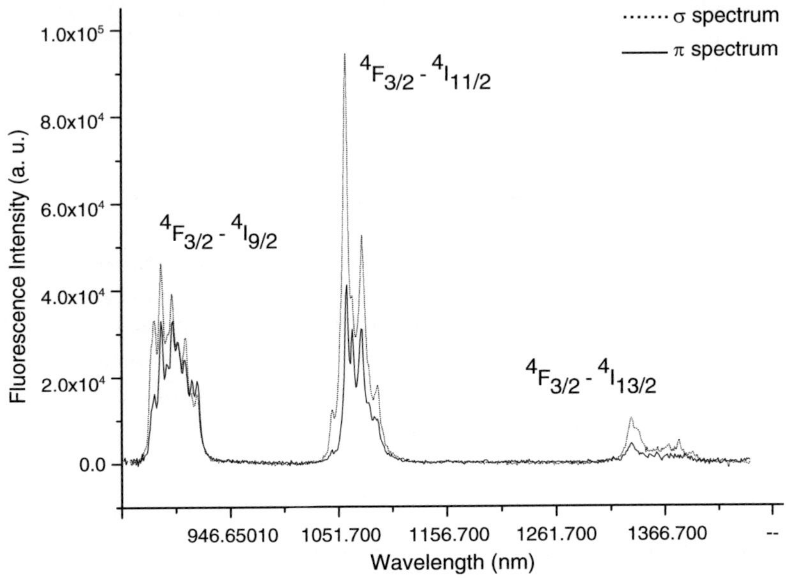

Fig. 22 Room temperature-polarized emission spectra of Nd^{3+}:$SrWO_4$ crystal

Table 10 Comparison of the spectroscopic parameters of Nd^{3+}:$SrWO_4$ crystal and other Nd^{3+}-doped tungstate crystals

Crystals	Ω_2	Ω_4	Ω_6	τ_f	τ_r	η (%)	σ_p	Ref.
Nd^{3+}:$SrWO_4$	46.35	12.46	13.71	198.26	188.80	95.23	1.75 (⊥c) 2.04(‖c)	[36]
Nd^{3+}:$La_2(WO_4)_2$	13.76	5.19	6.67	112	129	87	11.2	[37]
Nd^{3+}:$PbWO_4$	7.35	3.15	3.06	170	198	85.8	2.8	[38]
Nd^{3+}:Pb.WO_4	7.13	3.35	2.69	175	188	93.1	4.5 (⊥c) 5.7(‖c)	[39]
Nd^{3+}:$KY(WO_4)_2$	8.80	3.11	3.16	154	196	78.6	5.4	[40]

Table 11 The calculated radiative transition rate, the branching ratios, and the radiative lifetime for different transition levels

Transitions	Wavelength (nm)	Stimulated emission section (cm^2)		A (s^{-1})	β	τ (μs)
		σ polarized	π polarized			
$^4F_{3/2} \rightarrow {}^4I_{9/2}$	890	9.37×10^{-21}	2.49×10^{-20}	1.31×10^3	0.261	198.26
$^4F_{3/2} \rightarrow {}^4I_{11/2}$	1,060	1.75×10^{-20}	2.04×10^{-20}	2.98×10^3	0.59	
$^4F_{3/2} \rightarrow {}^4I_{13/2}$	1,334	3.84×10^{-20}	4.79×10^{-20}	713.35	0.14	
$^4F_{3/2} \rightarrow {}^4I_{15/2}$	1,852	–	–	38.09	7.55×10^{-3}	

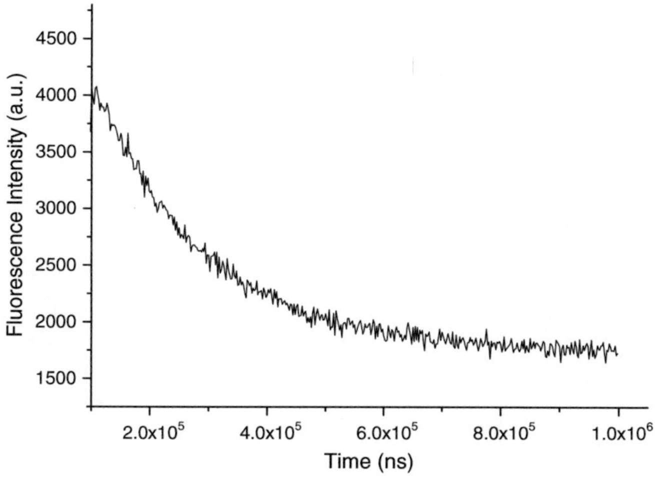

Fig. 23 Room temperature fluorescence decay curve of Nd^{3+}:$SrWO_4$ crystal

3.3.3 The Spectrum Characteristics of Tm^{3+}:$SrWO_4$ Crystal

The room temperature-polarized absorption spectra of Tm^{3+}:$SrWO_4$ crystal is presented in Fig. 24a [32], in which there are three strong absorption peaks centering at 1,744 nm, 1,213 nm,795 nm, 691 nm,473 nm, and 358 nm corresponding to the transitions from $^3H_6$ to $^3F_4$, $^3H_5$, $^3H_4$, $^3F_3$ + $^3F_2$, $^1G_4$, and $^1D_2$, respectively. The cross sections of the absorption peak for σ and π polarizations at 800 nm are 12 and 10.2, respectively. The FWHMs of the absorption bands for σ and π polarizations at 800 nm are 8 and 10 nm, respectively. Table 12 presents the comparison between the FWHM and absorption cross sections in Tm^{3+} doped crystal.

The room temperature-polarized emission spectra of Tm^{3+}:$SrWO_4$ crystal are presented in Fig. 25. The emission cross sections for σ and π polarizations at 1,500 nm corresponding to $^3H_4 \rightarrow {}^3F_4$ transition are 3.55×10^{-21} cm^2 and 4.22×10^{-21} cm^2, respectively. Table 13 presents the calculated radiative transition rate, the branching ratios, and the radiative lifetime for different transition levels of Tm^{3+}:$SrWO_4$ crystal. Figure 24b presents the room temperature fluorescence decay curve for $^3H_4 \rightarrow {}^3F_4$ transition of Tm $^{3+}$:$SrWO_4$ crystal. The lifetime is about 245.10 μs and relative luminescent quantum efficiency is about 69.84%.

3.3.4 The Spectrum Characteristics of Er^{3+}:$SrWO_4$ Crystal

The room temperature-polarized absorption spectra of Er^{3+}:$SrWO_4$ crystal are presented in Fig. 26 [46], in which there are three strong absorption peaks centering at 1,534 nm, 982 nm, 802 nm, 656 nm, 521 nm, 489 nm, 452 nm, 407 nm, and 380 nm corresponding to the transitions from $^4I_{15/2}$ to $^4I_{13/2}$, $^4I_{11/2}$, $^4I_{9/2}$, $^4F_{9/2}$, $^2H_{11/2}$, $^4F_{7/2}$,

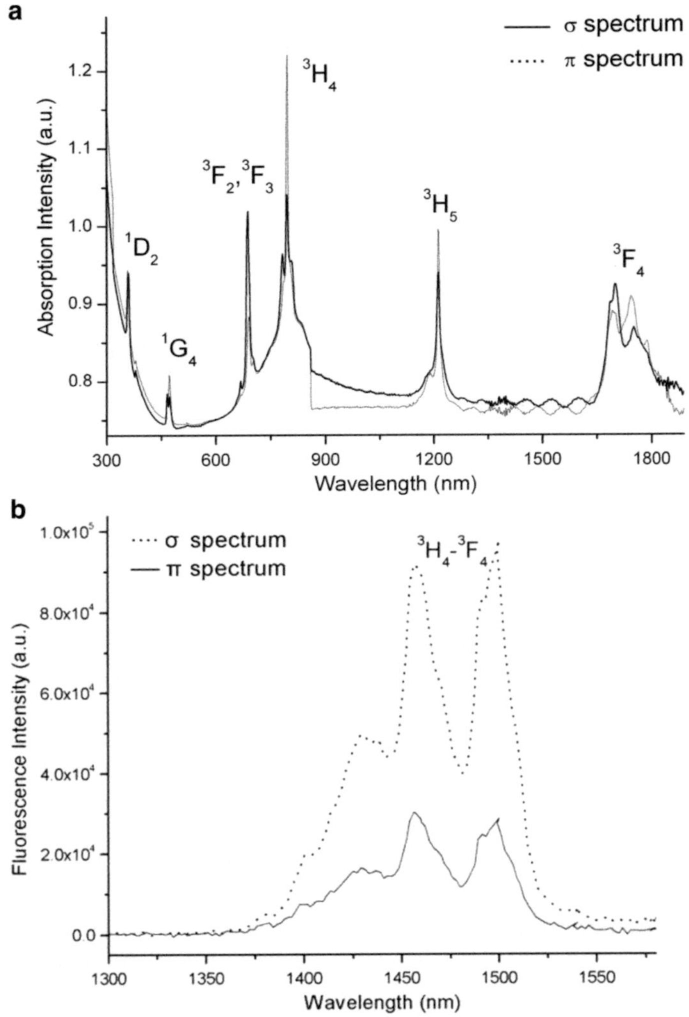

Fig. 24 (**a**) Polarized room temperature absorption spectra of Tm^{3+}:$SrWO_4$ crystal. (**b**) Polarized room temperature fluorescence of Tm^{3+}:$SrWO_4$ crystal

$^4F_{3/2}$, $^4F_{5/2}$, $(^4G, {}^4F, {}^2H)_{9/2}$ and $^2G_{7/2}, {}^4G_{9/2}, {}^2K_{15/2}, {}^4G_{11/2}$, respectively. The room temperature-polarized emission spectra of Er^{3+}:$SrWO_4$ crystal is presented in Fig. 27. The emission cross sections for σ and π polarizations at 1,500 nm corresponding to $^3H_4 \rightarrow {}^3F_4$ transition are 3.55×10^{-21} cm^2 and 4.22×10^{-21} cm^2, respectively. Table 14 shows the calculated radiative transition rate, the branching ratios, and the radiative lifetime for different transition levels. Figure 28 presents the room temperature fluorescence decay curve for $^4I_{13/2} \rightarrow {}^4I_{15/2}$ transition of Er^{3+}:$SrWO_4$ crystal. The lifetime is about 5,755 μs and relative luminescent quantum efficiency is about 18.52%.

Table 12 Comparison between the FWHM and absorption cross section in Tm^{3+}-doped crystal

Crystals	Polarization spectrum	Wavelength (nm)	FWHM (nm)	Cross section (10^{-20} cm^2)	Ref.
5 at.% $Tm:YVO_4$	π	797.5	5	2.5	[42]
0.5 at.% $Tm:LiTaO_3$	σ	795	9.5	6	[43]
0.7 at.% $Tm:KYb(WO_4)_2$	π	793.6	3.5	8.8	[44]
	σ	801.8	1.4	4.8	
2.02 at.% $Tm:GdAl_3(BO_3)_4$	π	797	15	1.21	[45]
	σ	810	13	2.98	
0.24 wt% $Tm^{3+}:SrWO_4$	π	795	8	12.0	[32]
	σ	794	10	10.2	

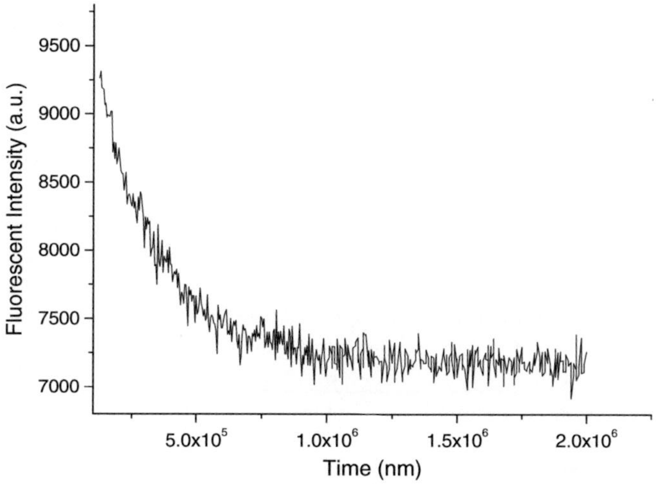

Fig. 25 Room temperature luminescence decay curves excited by 805 nm

Table 13 The calculated radiative transition rate, the branching ratios, and the radiative lifetime for different transition levels of $Tm^{3+}:SrWO_4$ crystal

Start levels	Terminal levels	Wavelength (nm)	A (s^{-1})	β (A)	τ (μs)
$^1D_2$	$^3H_6$	359	5,440	0.089	16.41
	$^3F_4$	450	47,840	0.785	
	$^3H_5$	508	212	0.003	
	$^3H_4$	652	3,266	0.054	
	$^3F_3$,$^3F_2$	758	3,778	0.062	
	$^1G_4$	1,491	398	0.007	
$^1G_4$	$^3H_6$	472	1,952	0.436	223.06
	$^3F_4$	644	285	0.064	
	$^3H_5$	770	1,583	0.353	
	$^3H_4$	1,158	567	0.126	
	$^3F_3$,$^3F_4$	1,543	96	0.021	
$^3H_4$	$^3H_6$	798	2,624	0.92	350.97
	$^3F_4$	1,452	215	0.075	
	$^3H_5$	2,300	10	0.004	
$^3F_4$	$^3H_6$	1,771	76	1	13,158

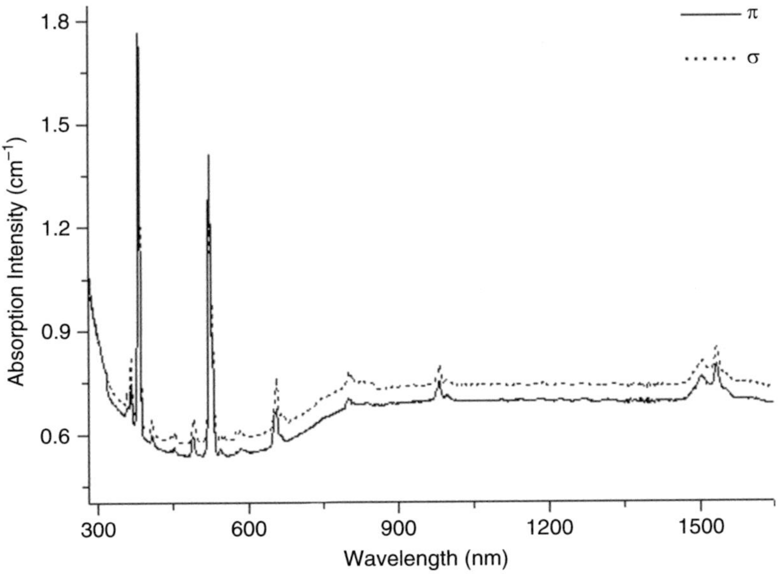

Fig. 26 Polarized room temperature absorption spectra of Er^{3+}:$SrWO_4$ crystal

Table 14 Calculated electric dipole and magnetic dipole transition probabilities, radiative lifetime, branching ratios, and experimental lifetimes of Er^{3+} in $SrWO_4$ crystal

$J \rightarrow J'$	Average frequency (cm^{-1})	β	A^{ed} (s^{-1})	A^{md} (s^{-1})	τ(ms)
$^4I_{13/2} \rightarrow {}^4I_{15/2}$	6,494	1.00	20.12	12.06	31.08
$^4I_{11/2} \rightarrow {}^4I_{15/2}$	10,225	0.877	540.84		1.622
$^4I_{11/2} \rightarrow {}^4I_{13/2}$	3,698	0.123	63.07	12.498	
$^4I_{9/2} \rightarrow {}^4I_{15/2}$	12,516	0.663	349.80		1.896
$^4I_{9/2} \rightarrow {}^4I_{13/2}$	5,989	0.323	170.49		
$^4I_{9/2} \rightarrow {}^4I_{11/2}$	2,291	0.014	3.05	4.12	
$^4F_{9/2} \rightarrow {}^4I_{15/2}$	15,337	0.897	4.595×10^3		0.195
$^4F_{9/2} \rightarrow {}^4I_{13/2}$	8,810	0.044	226.707		
$^4F_{9/2} \rightarrow {}^4I_{11/2}$	5,112	0.057	245.71	44.49	
$^4F_{9/2} \rightarrow {}^4I_{9/2}$	2,821	2.62×10^{-3}	9.79	3.61	
$^4S_{3/2} \rightarrow {}^4I_{15/2}$	18,382	0.681	4.68×10^3		0.146
$^4S_{3/2} \rightarrow {}^4I_{13/2}$	11,855	0.264	1.81×10^3		
$^4S_{3/2} \rightarrow {}^4I_{11/2}$	8,157	0.025	174.55		
$^4S_{3/2} \rightarrow {}^4I_{9/2}$	5,866	0.029	200.10		

3.4 The Raman and SRS Laser Characteristics

3.4.1 The Stimulated Raman Scattering Self-Frequency Conversion

Laser pulses were generated at 1,057 nm corresponding to the $^4F_{3/2} \rightarrow {}^4I_{11/2}$ transition and SRS self-frequency conversion pulses were observed at 1,171 nm (Stokes 1) in *c*-polarization and the maximum Stokes 1 SRS conversion was 1.8% [47].

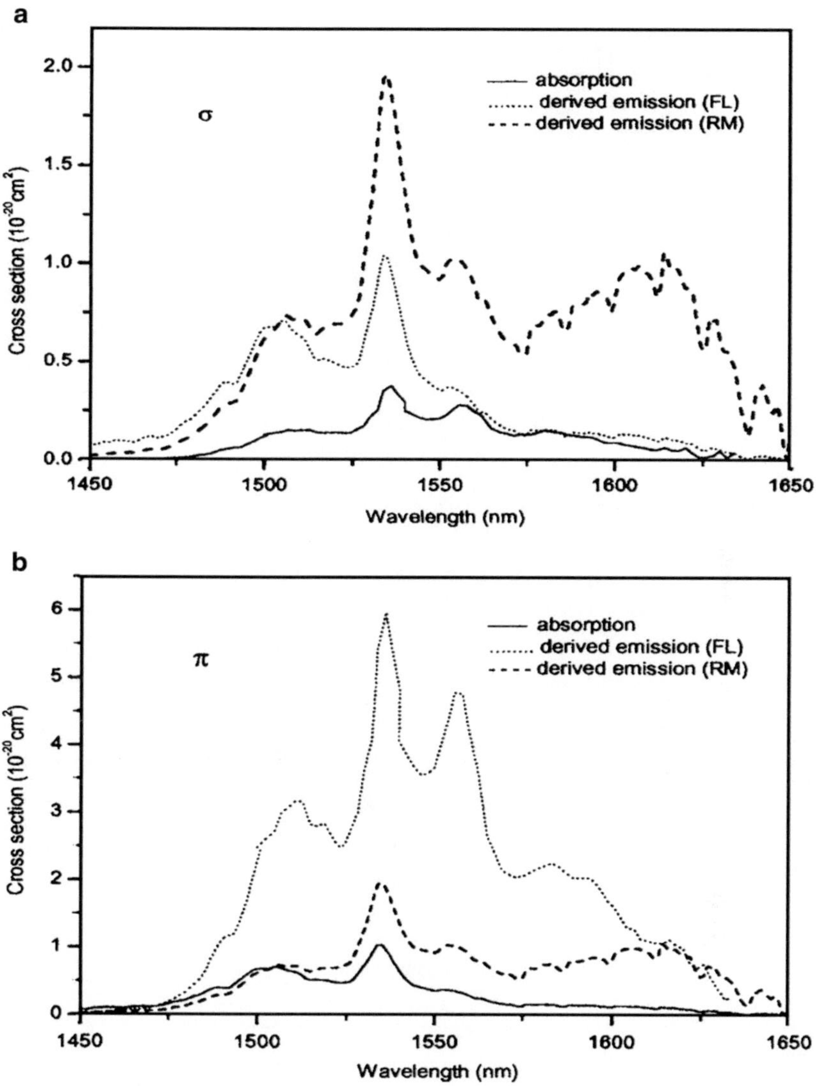

Fig. 27 Polarized room temperature fluorescence spectra of Er^{3+}:$SrWO_4$ crystal

Figure 29 presents the laser and Stokes 1 output power from self-stimulated Raman scattering of the ${}_{4}F_{3/2} \rightarrow {}^{4}I_{11/2}$ laser transition. Figure 30 shows the time evolutions of the pump, laser, and Stokes 1 for ${}^{4}F_{3/2} \rightarrow {}^{4}I_{11/2}$ transition.

The ${}^{4}F_{3/2} \rightarrow {}^{4}I_{13/2}$ transition-generated laser pluses at 1,334 nm and SRS Stokes 1 at 1,517 nm, the latter is close to the eye-safe laser. The maximum conversion efficiency of the pump toward the Stokes 1 SRS was 0.4%. Figure 31 presents the Stokes 1 output power from self-stimulated Raman scattering of the ${}^{4}F_{3/2} \rightarrow {}^{4}I_{13/2}$ laser

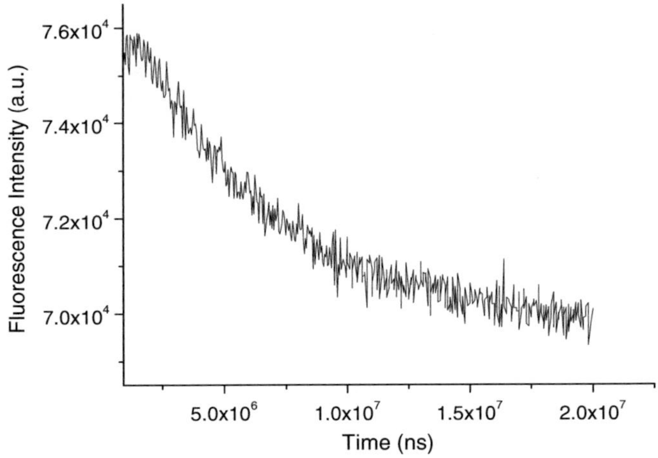

Fig. 28 Room temperature luminescence decay curves excited by 521 nm

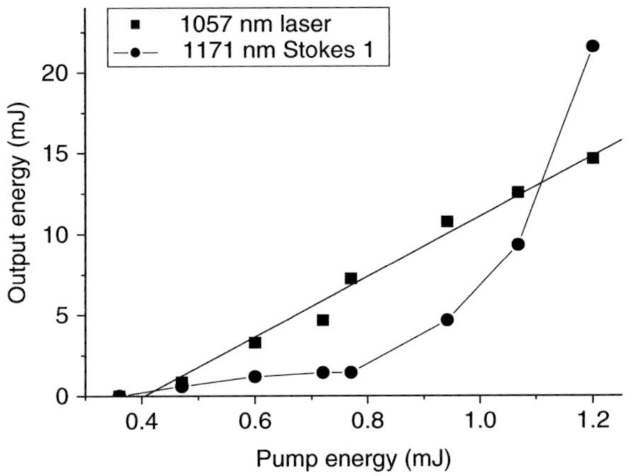

Fig. 29 Laser and Stokes 1 output power from self-stimulated Raman scattering of the $^4F_{3/2} \rightarrow {}^4I_{11/2}$ laser transition

transition. Figure 32 shows the Raman spectra of $SrWO_4$:Nd^{3+} crystal relevant to self-stimulated Raman scattering.

3.4.2 The Continuous-Wave Nd:YVO_4–$SrWO_4$ Intracavity Raman Laser at 1,179.5 nm

A continuous-wave intracavity Raman laser at 1,179.5 nm with a $SrWO_4$ Raman crystal in a diode-end-pumped Nd:YVO_4 laser was reached [48]. The highest output power of 2.23 W is obtained at the laser diode power of 21.2 W corresponding to the

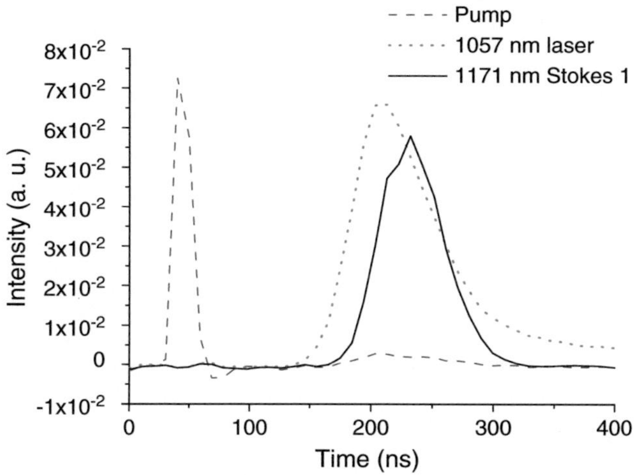

Fig. 30 Time evolutions of the pump, laser, and Stokes 1 for $^4F_{3/2} \rightarrow {}^4I_{11/2}$ transition

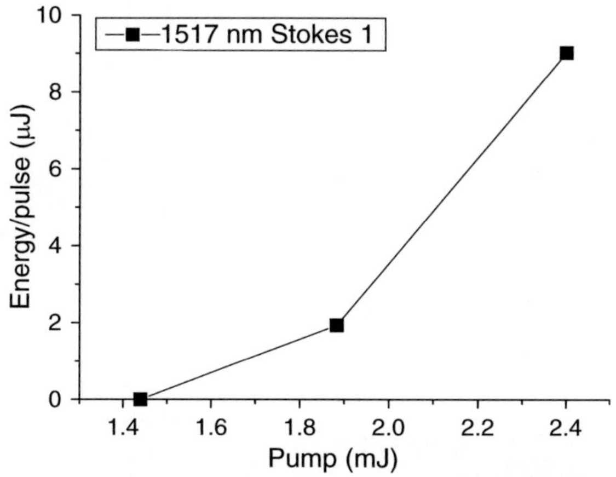

Fig. 31 Stokes 1 output power from self-stimulated Raman scattering of the $^4F_{3/2} \rightarrow {}^4I_{13/2}$ laser transition

slope efficiency of 10.5%. The measured Raman thresholds are about 9.3 W and 8.3 W in the diode pump laser power for the *a*- and *b*-polarized configurations, respectively. The Raman gain coefficients of the c-cut $SrWO_4$ Raman crystal for the *a*- and *b*-polarized pumps are estimated to be about 4.9 cm/GW and 4.7 cm/GW, respectively. Figure 33 presents the experimental arrangement of the cw diode-end-pumped Nd:YVO_4/$SrWO_4$ intracavity Raman laser. Figure 34 shows the output power at 1,064 and 1,179.5 nm as functions of incident pump power at 808 nm.

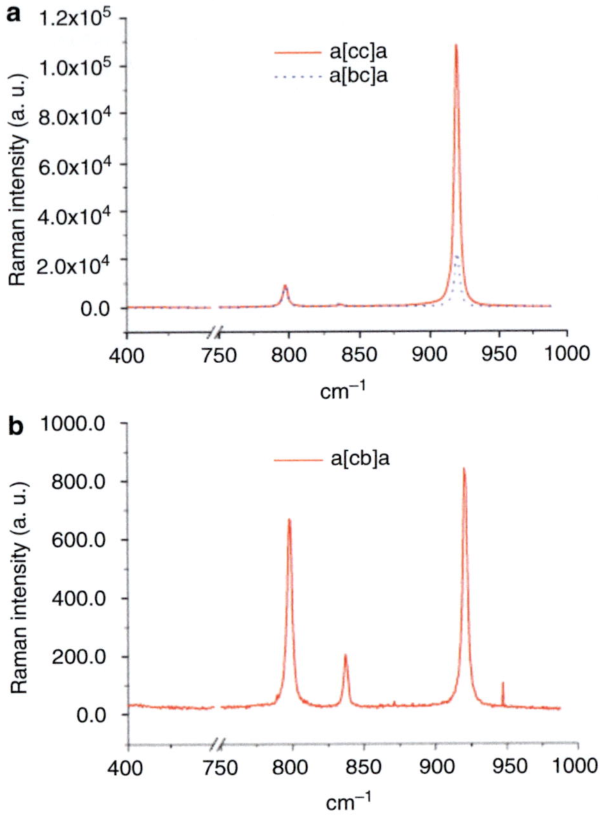

Fig. 32 Raman spectra of $SrWO_4:Nd^{3+}$ crystal relevant to self-stimulated Raman scattering

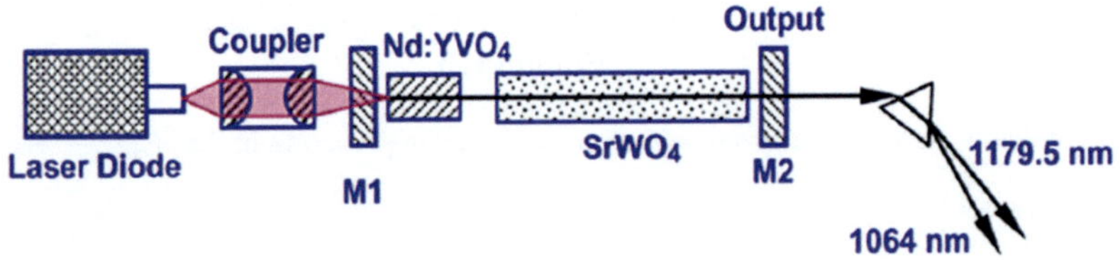

Fig. 33 The experimental arrangement of the cw diode-end-pumped $Nd:YVO_4/SrWO_4$ intracavity Raman laser

Figure 35 presents the Raman conversion efficiencies versus the incident pump power in the *a*- and *b*-polarized configurations. Figure 36 shows the output spectrum of the cw $Nd:YVO_4/SrWO_4$ intracavity Raman laser.

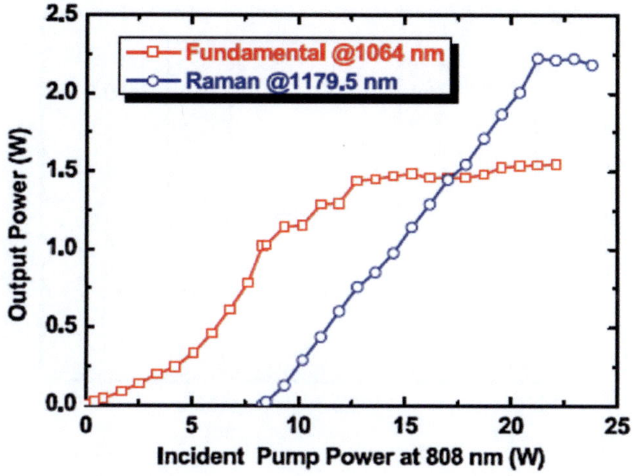

Fig. 34 The output power at 1,064 and 1,179.5 nm as functions of incident pump power at 808 nm

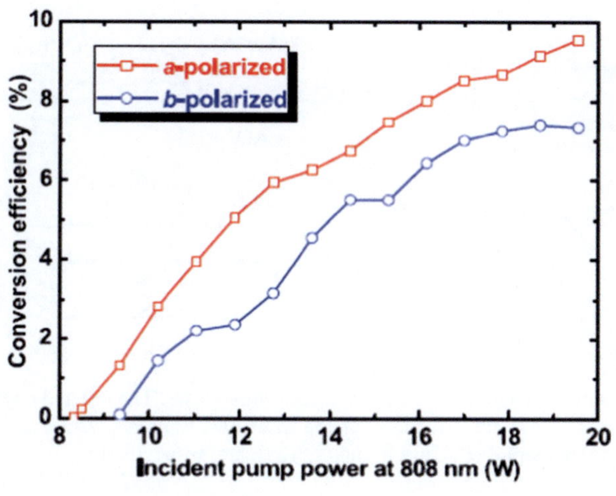

Fig. 35 The Raman conversion efficiencies versus the incident pump power in the *a*- and *b*-polarized configurations

3.4.3 The High-Efficiency $Nd:YVO_4$–$SrWO_4$ Eye-Safe Raman Laser at 1,531 nm

A high-efficiency diode-end-pumped Q-switched eye-safe linearly polarized intracavity Raman laser at 1,531 nm is demonstrated, with $Nd:YVO_4$ as the laser medium and $SrWO_4$ as the Raman crystal [49]. The highest average power of 1.93 W is achieved, with an incident pump of 15.6 W and a repetition rate of 35 KHz. The narrowest pulse duration of 4.9 ns and the highest peak power of

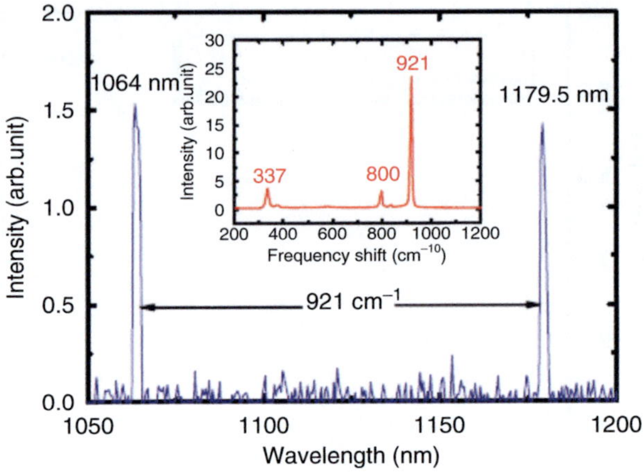

Fig. 36 The output spectrum of the cw Nd:YVO_4/$SrWO_4$ intracavity Raman laser. The inset is the spontaneous Raman spectrum of the $SrWO_4$ crystal

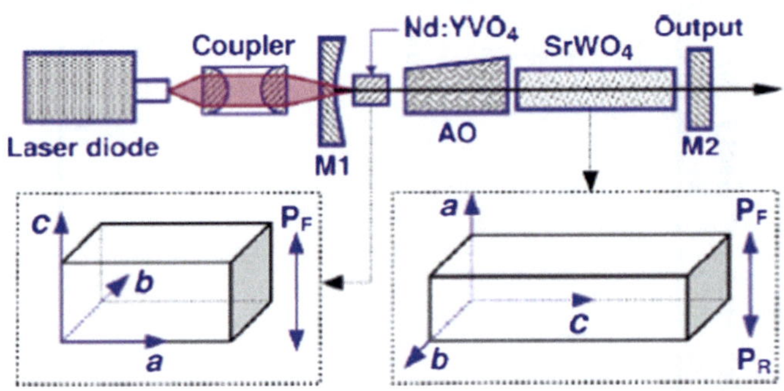

Fig. 37 Experiment arrangement of an intracavity Raman laser. The *left* and *right* insets show the schematic diagrams of the cut directions of the laser and Raman crystals and the polarization directions of the fundamental (P_F) and Raman (P_R) lasers, respectively

32.2 KW are obtained at a repetition rate of 5 KHz. Figure 37 shows the experiment arrangement of an intracavity Raman laser. Figure 38 presents the average output power at 1,531 nm as a function of incident pump power at 808 nm at repetition rate of 35 KHz for four output couples. Figure 39 also presents the average output power at 1,531 nm as a function of incident pump power at 808 nm, with the output coupler of 30%, at four repetition rates. Figure 40 shows the dependence of pulse width at 1,531 nm on incident pump power at four repetition rates. Figure 41 shows the pulse profiles of the Raman and fundamental pulses at a repetition rate of 5 KHz.

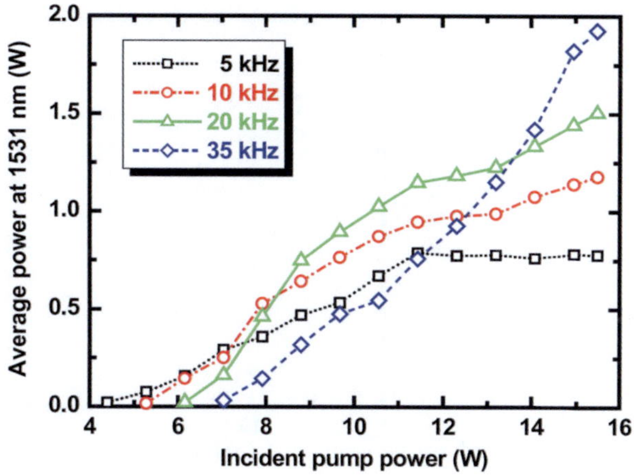

Fig. 38 Average output power at 1,531 nm as a function of incident pump power at 808 nm at repetition rate of 35 KHz for four output couples

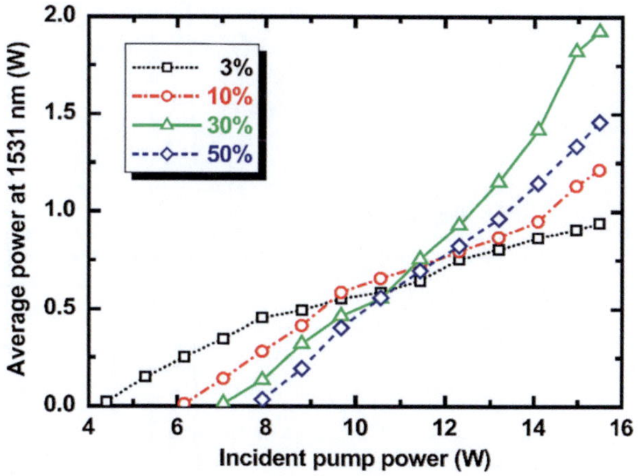

Fig. 39 Average output power at 1,531 nm as a function of incident pump power at 808 nm, with the output coupler of 30% at four repetition rates

3.4.4 Highly Efficient Nd:YAG–$SrWO_4$ Intracavity Raman Laser at 1,180 nm

A highly efficient diode-pumped actively Q-switched intracavity Raman laser with SrWO4 as the Raman-active medium is presented. As high as 23.8% diode-to-Stokes optical conversion efficiency is obtained with an incident pump power of 7.17 W and a pulse repetition rate of 15 KHz [50]. Figure 42 shows the schematic of the experimental setup. Figure 43 presents the average output power at 1,064 nm as a function of incident pump power at 808 nm for the PRFs of 10, 15, and 20 KHz. Figure 44 presents the average output power at the Stokes wavelength as a function of incident pump power for the PRFs of 10, 15, and 20 KHz. Figure 45 shows

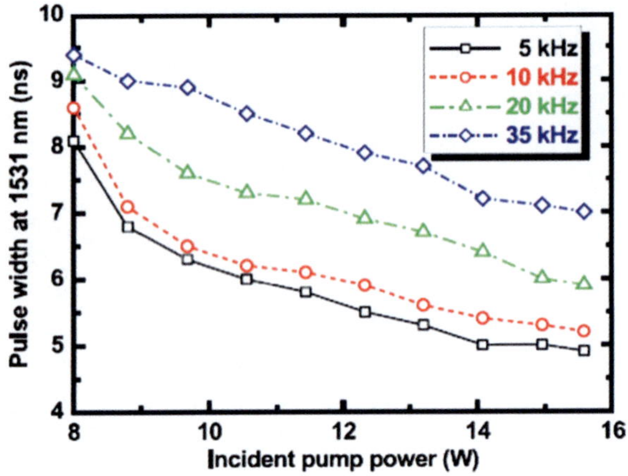

Fig. 40 Dependence of pulse width at 1,531 nm on incident pump power at four repetition rates

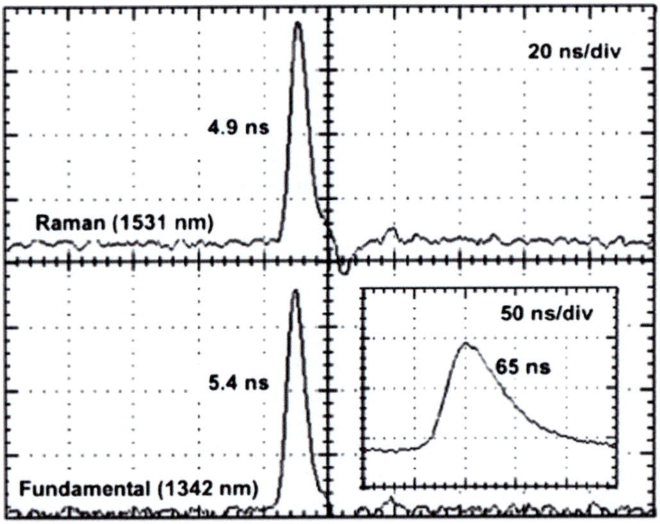

Fig. 41 Pulse profiles of the Raman and fundamental pulses at a repetition rate of 5 KHz. Inset is the pulse profile of the fundamental when the Raman crystal is removed at repetition rate of 35 KHz

the pulse width at the Stokes wavelength as a function of incident pump power for the PRFs of 10, 15, and 20 KHz. Figure 46 shows the typical oscilloscope traces for the fundamental and Raman pulses. Figure 47 shows the optical spectra for the actively Q-switched intracavity Nd:YAG–SrWO4 Raman laser.

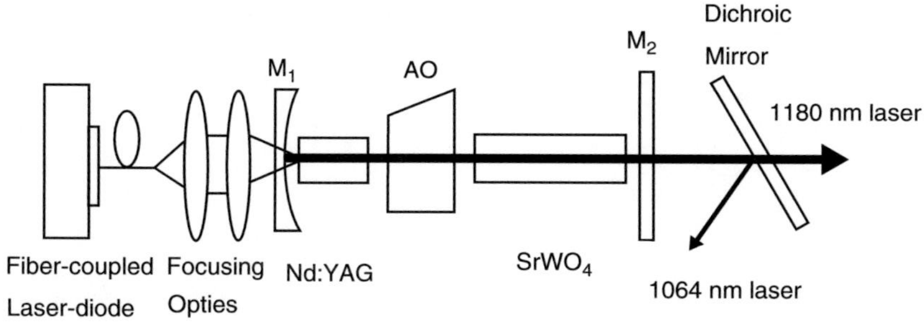

Fig. 42 Schematic of the experimental setup

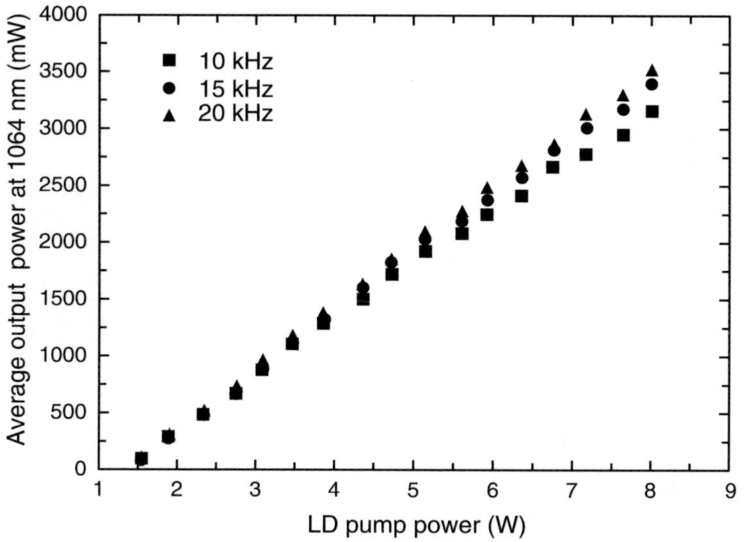

Fig. 43 Average output power at 1,064 nm as a function of incident pump power at 808 nm for the PRFs of 10, 15, and 20 KHz

3.4.5 The Efficient Nd:YAG/$SrWO_4$/KTP Yellow Laser

An efficient intracavity frequency-doubled Raman laser was obtained by Xingyu Zhang using a $SrWO_4$ Raman medium, an Nd:YAG ceramic gain medium, and a KTP frequency-doubled medium, three laser cavities, including a two-mirror cavity, a three-mirror coupled cavity, and a folded cavity, were investigated [51]. With the coupled cavity, a 2.93 W, 590 nm laser was obtained at an incident pump power of 16.2 W and a pulse repetition frequency of 20 KHz; the corresponding conversion efficiency was 18.1%. The highest conversion efficiency of 19.2% was obtained at an incident pump power of 14.1 W and a pulse repetition frequency of 15 KHz. The obtained maximum output power and conversion efficiency were

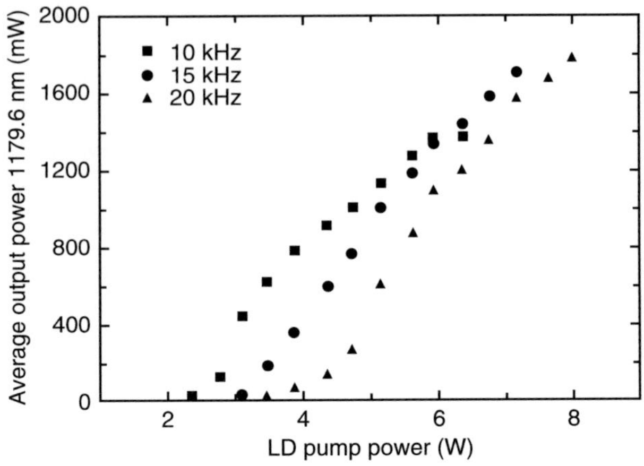

Fig. 44 Average output power at the Stokes wavelength as a function of incident pump power for the PRFs of 10, 15, and 20 KHz

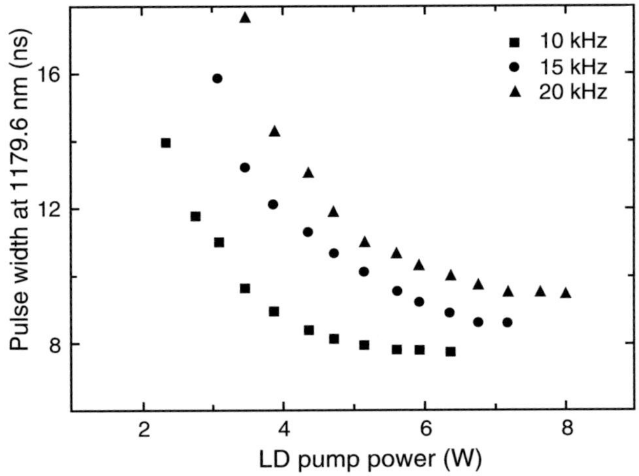

Fig. 45 Pulse width at the Stokes wavelength as a function of incident pump power for the PRFs of 10, 15, and 20 KHz

much higher than the results previously obtained with intracavity frequency-doubled solid-state Raman lasers.

An efficient yellow–green laser at 559.6 nm based on the sum–frequency generation of diode-end-pumped Nd:YAG/$SrWO_4$ Raman laser is also demonstrated by Ge Zhang [52]. The 45 nm in length $SrWO_4$ crystal grown by Czochralski method and the KTP crystal with a type-II critical phase-matching ($\theta = 83.4°$, $\Phi = 0°$) cut were adopted for Raman conversion and sum–frequency generation, respectively. The maximum average power of 2.41 W was achieved at the incident pump power of

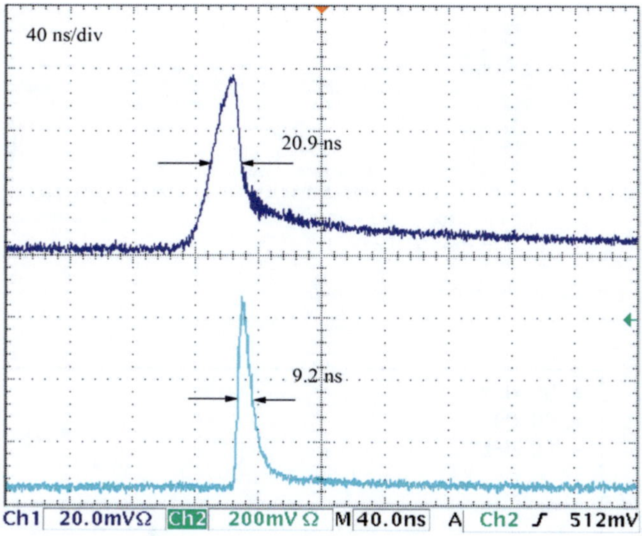

Fig. 46 Typical oscilloscope traces for the fundamental and Raman pulses

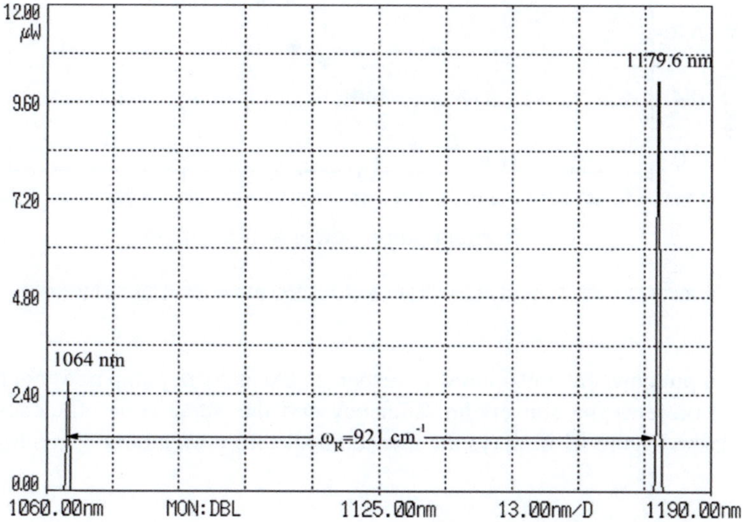

Fig. 47 Optical spectra for the actively Q-switched intracavity Nd:YAG-SrWO4 Raman laser

20.4 W and the pulse repetition rate of 60 KHz. The overall diode-yellow conversion efficiency was 11.8% and the slope efficiency was about 24%. The results show that the critical pump power of resonator instability is increased with the pulse repetition rate. Figure 48 shows the schematic of the experimental setup. Figure 49 presents the

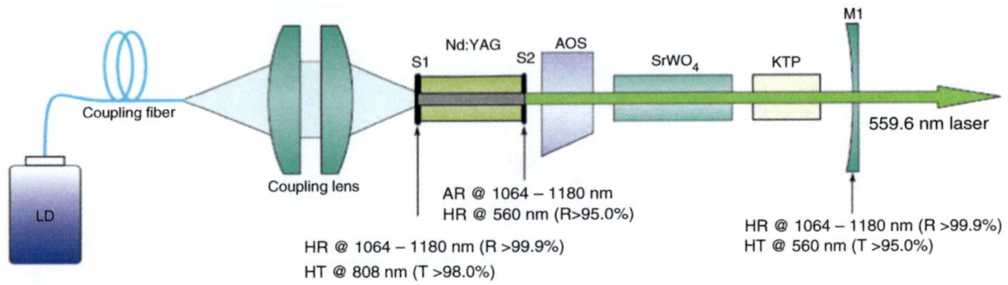

Fig. 48 The schematic of the experimental setup

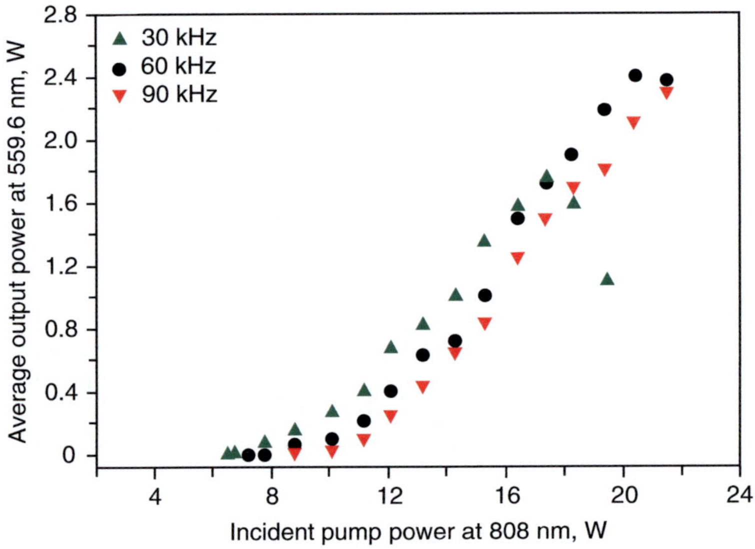

Fig. 49 The average output power at 559.6 nm with respect to the incident pump power at 808 nm

average output power at 559.6 nm with respect to the incident pump power at 808 nm. Figure 50 presents the conversion efficiency and the fitted slope efficiency at the 590 nm laser. Figure 51 displays the measured spectrum of yellow–green light.

3.4.6 The Continuous-Wave Nd:YVO_4/$SrWO_4$/LBO Yellow Laser

An end-pumped continuous-wave intracavity yellow Raman laser at 590 nm with a 60 mm long pure crystal $SrWO_4$ and an intracavity LiB_3O_5 frequency doubling crystal was demonstrated by Fugui Yang [53]. The highest output power of yellow laser at 590 nm was 230 MW and the output power and threshold were found to be correlative with the polarized directions of pure single crystal $SrWO_4$ deeply. Along the different directions, the minimum and maximum thresholds of yellow

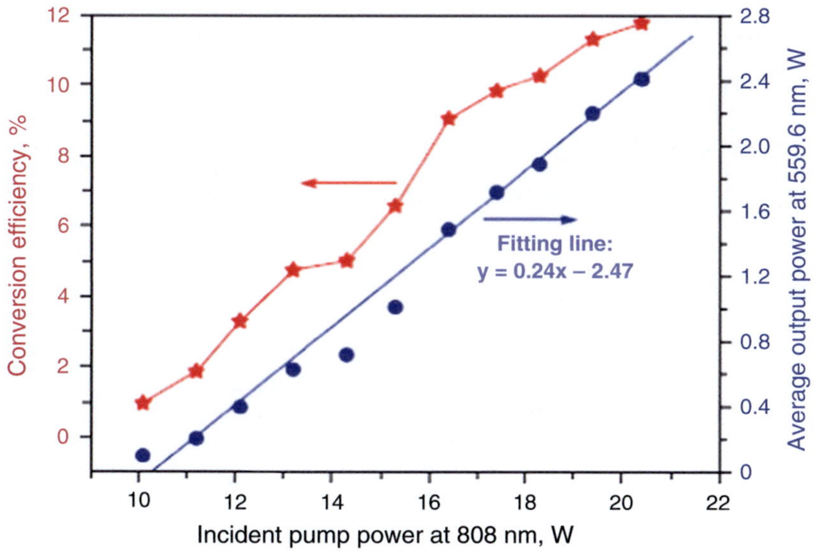

Fig. 50 The conversion efficiency and the fitted slope efficiency at the 590 nm laser

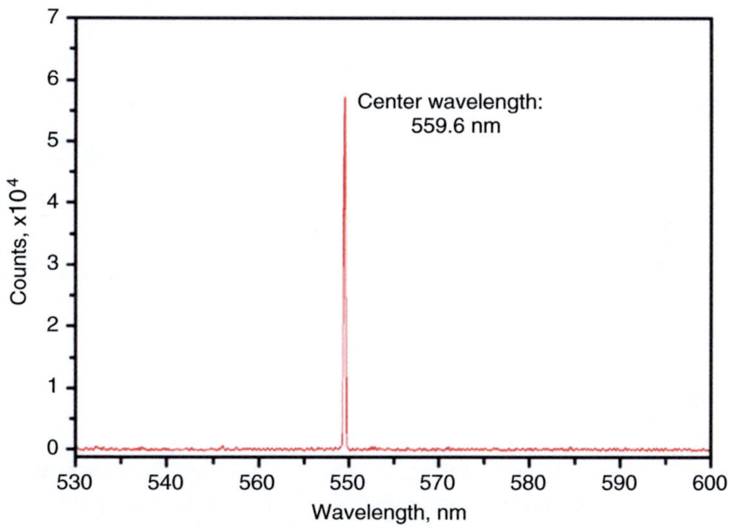

Fig. 51 The measured spectrum of *yellow–green* light

Raman laser at 590 nm were measured to be 2.8 and 14.3 W with respect to 808 nm LD pump power, respectively. Figure 52 shows the schematic of the experimental setup. Figure 53 presents the average output power at 559.6 nm with respect to the incident pump power at 808 nm.

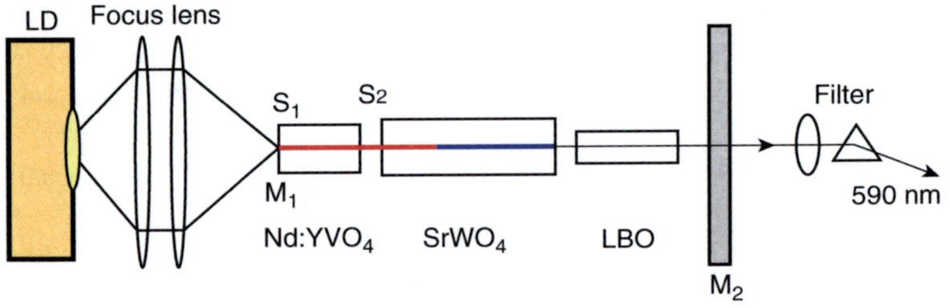

Fig. 52 Schematic of the experimental setup

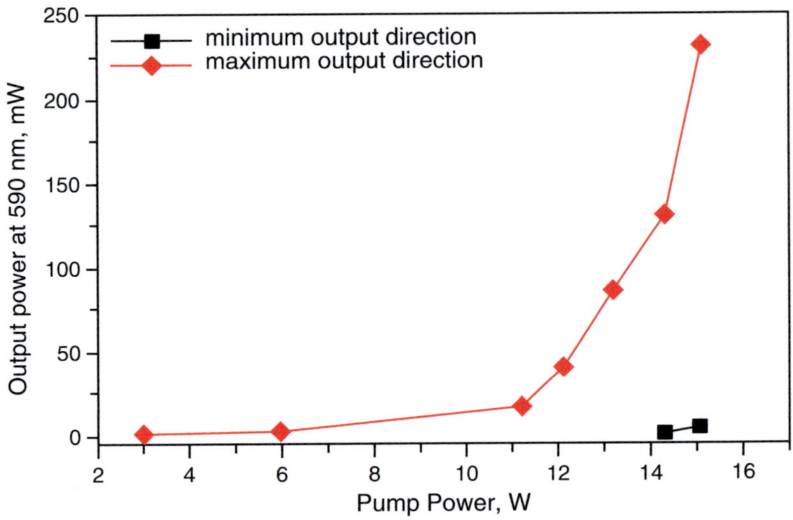

Fig. 53 Average output power at 559.6 nm with respect to the incident pump power at 808 nm

4 Summary

The growth, PL spectra, and Raman spectra, stimulated Raman scattering self-frequency conversion laser characteristics of pure and rare earth-doped KGd $(WO_4)_2$, $SrWO_4$ crystals were reviewed. The SRS self-frequency conversion laser technology was dealt with. As a result, the self-Raman laser outputs with high efficiency at ~1,180 nm wavelength and its frequency-doubling laser outputs at ~590 nm wavelength achieved have been presented. The crystal of $KGd(WO_4)_2$ is easier to accept rare earth doping ion, whereas $SrWO_4$ crystal has higher Raman plus coefficient and is easier to be grown with higher optical quality and large sizes.

Acknowledgments Some works of this chapter were supported by National Nature Science Foundation of China (No.50902129, 61078076, 91122033), Major Projects from FJIRSM (SZD09001), the Knowledge Innovation Program of the Chinese Academy of Sciences (Grant No. KJCX2-EW-H03), Science and Technology Plan Major Project of Fujian Province of China (Grant No. 2010I0015).

References

1. Hellwarth RW (1963) Phys Rev 130:1850
2. Sen PK, Sen P (1986) Phys Rev B 33:5038
3. Malkin VM, Tsidulko YA, Fisch NJ (2000) Phys Rev Lett 85:4068
4. Li Z, Li L, Tian H, Zhou G, Spatschek KH (2002) Phys Rev Lett 89:263901
5. Takahashi J, Matsubara E, Arima T, Hanamura E (2003) Phys Rev B 68:155102
6. Shen YR (2003) The principles of nonlinear optics. Wiley-Interscience, Hoboken
7. Levenson MD (1974) IEEEE J Quant Electron QE 10:110
8. Harris SE, Sokolo AV (1997) Phys Rev A 55:R4019
9. Zverev PG, Basiev TT, Prokhorov AM (1999) Opt Mater 11:335
10. Kaminskii AA, Eichler HJ, Grebe D et al (1998) Opt Mater 10:269
11. Qi X, Luo Z, Liang J (2000) J Cryst Growth 246:363–366
12. Kaminskii A, Ueda K, Eichler HJ, Lu J et al (2001) Opt Comm 194:201
13. Chen W, Inagawa Y, Omatsu T et al (2001) Opt Comm 194:401
14. Kaminskii A, Sarkiso SE, Pavlyuk A, Lyubchenko V (1980) Izv Akad Nauk SSSR Neorg Mater 16:501
15. Jia G (2005) The study on the growth, structure, spectral, and laser properties of $Tm_xGd_{1-x}Al(BO_3)_4$ and rare-earth doped $SrWO_4$ crystals. Master Dissertation, Graduated School of Chinese Academy of Sciences
16. Tu C, Luo Z, Chen G, Zhao T (1995) J Cryst Growth 152(3):235–237
17. Johnson LF, Dietz RE, Guggenheim HJ (1963) Phys Rev Lett 11:318–320
18. Tu C, Li J, Zhu Z et al (2003) J Cryst Growth 256(1–2):63–66
19. Tu C, Li J, You Z et al (2004) Chin Laser 31:377
20. Tu C (2005) The study on the growth, structure, spectra and laser characteristics of new rare earth-actived laser crystals. Doctor Dissertation, Graduated School of Chinese Academy of Sciences
21. Yingwei W, Haobo C, Jinghe L et al (2004) Opt Tech (Chinese) 30(6):717
22. Brenier A, Tu C, Li J et al (2001) J Phys Condens Matter 13:4097–4103
23. Zundu Luo, Xueyuan Chen, Tu C (1997) Acta Optica Sinica (Chinese) 17(8):1144
24. Tu C, Li J, Zhaojie Z et al (2003) Opt Comm 227(4–6):383–388
25. Huang Jianhong, Lin Jipeng, Su Rongbing et al (2007) Opt Lett 32(9):1096
26. Gurmen E, Daniels E, King JS (1971) J Chem Phys 55:1093–1097
27. Ivleva LI, Basiev TT, Voronina IS et al (2003) Opt Mater 23:439
28. Sattler JP, Nemarich J (1970) Phys Rev B1:4249
29. Choosuwan H, Guo R, Bhalla AS et al (2002) J Appl Phys 91:5051
30. Carvajal JJ, Sole R, Gavalsa J et al (2003) Chem Matter 15:2730
31. Chauhan AK (2003) J Cryst Growth 254:418
32. Jia G, Tu C, You Z et al (2005) Solid State Comm 134(9):583–588
33. Brixner LH, Sleight AW (1973) Mater Res Bull 8:1269
34. Gongming Wang, Wencheng Wang (1982), Physics 11(3):164
35. Born M, Wolf E (1975) Principles of optics. Pergamon, Oxford
36. Jia G, Tu C, Brenier A (2005) Appl Phys B 81:627–633
37. Métrat G, Muhlstein N, Brenier A et al (1997) Opt Matter 8:75

38. Zubenko DA, Noginov MA, Semenkov SG et al (1992) Sov J Quant Electron 22:133
39. Weber MJ, Varitimos TE (1971) J Appl Phys 42:4996
40. Chen Y, Lin X, Luo Z et al (2003) Chem Phys Lett 381(5–6):598
41. Chen W, Inagawa Y, Omatsu T et al (2001) Opt Comm 194:201
42. Ohta K, Saito H, Obara M (1993) J Appl Phys 73:3149
43. Sokolska I, Ryba-Romanowski W, Golab S et al (2000) J Chem Sol 61:1573
44. Pujol MC, Guell F, Mateos X et al (2002) Phys Rev B 66:144304
45. Jia GH, Tu CY, Li JF et al (2004) J Appl Phys 96:6262
46. Jia G, Tu C, Zhenyu You et al (2005) J Appl Phys 98:093525
47. Brenier A, Jia G, Tu C (2004) J Phys Condens Matter 16:9103–9108
48. Fan L, Fan YX, Duan YH et al (2009) Appl Phys B 94:553–557
49. Fan YX, Liu Y, Duan YH et al (2008) Appl Phys B 93:327–330
50. Chen X, Zhang X, Wang Q et al (2008) Opt Lett 33(7):705–707
51. Zhenhua C, Xingyu Z, Qingpu W, Liu Z et al (2009) Opt Lett 34(17):2610–2612
52. Duan YM, Zhu HY, Zhang G, Huang CH, Wei Y, Tu CY, Zhu ZJ, Yang FG, You ZY (2010) Laser Phys Lett 7(7):491–494
53. Yang FG, You ZY, Zhu ZJ et al (2010) Laser Phys Lett 7(1):14–16

Struct Bond (2012) 145: 121–134
DOI: 10.1007/430_2011_74

Published online: 3 March 2012

Exploration of New Second-Order Nonlinear Optical Compounds Containing Main Group Elements

Hua-Jun Zhao, Xin-Tao Wu, and Li-Ming Wu

Abstract This review mainly highlights recent research progress on the exploration syntheses, crystal structures, and nonlinear optical properties of multinary chalcogenides. Some examples show that slight radius change (Ga^{3+} vs. In^{3+}) leads to different packing patterns of the same asymmetric units that eventually result in NLO properties with different origins. Besides, combination of two types of asymmetric units, SbQ_x polyhedron and TrQ_4 tetrahedron, not only gives rise to a rich structural chemistry but also yields many NCS compounds with excellent IR NLO performance. Additionally, the studies of the new mid-IR NLO crystal $BaGa_4S_7$ developed by a Bridgman–Stockbarger technique and a series new NLO chalcogenides based on TrQ_4 and TtQ_4 tetrahedra are also included.

Keywords Antimony · Asymmetric unit · Chalcogenide · Main group element

Contents

H.-J. Zhao, X.-T. Wu, and L.-M. Wu (✉)
State Key Laboratory of Structural Chemistry, Fujian Institute of Research on the Structure of Matter, Chinese Academy of Sciences, Fuzhou 350002, People's Republic of China
e-mail: Liming_wu@fjirsm.ac.cn

1 Introduction

Noncentrosymmetric (NCS) compounds are of current interest due to their technologically important properties such as second-harmonic generation (SHG) [1]. The development of numerous nonlinear optical (NLO) crystals such as BaB_2O_4 (BBO) [2], LiB_3O_5 (LBO) [3], KH_2PO_4 (KDP) [4], $KTiOPO_4$ (KTP) [5, 6], $AgGaQ_2$ (Q = S, Se), and $ZnGeP_2$ (ZGP) [7–11] has resulted in significant advances in laser device application from ultraviolet to infrared regions. Compared to the oxide-based NLO materials, chalcogenides possess higher SHG coefficients owing to the highly polarizable M–S (Se) bonds and wider transparent window in the IR region, yet their lower laser-damage threshold caused by the relatively small bandgaps limited their applications [12]. Therefore, the explorations for new material systems with excellent IR NLO performance are of broad interest in NLO material science and laser technology.

The involvement of asymmetric building unit is an effective approach to generate NCS structures that possibly exhibit good SHG properties. Considerable compounds containing Se^{4+}, Te^{4+} with stereochemically active lone pair electrons, as well as d^0 transition metal (Ti^{4+}, Nb^{5+}, W^{6+}, etc.) in a distorted octahedral coordination environment have shown excellent SHG properties [13–18].

SbQ_x is also an asymmetric coordination sphere as a consequence of stereochemically active lone pair electrons of antimony [19–22]. However, the number of NLO-active antimony chalcogenides is still rare, such as Ag_3SbS_3 [23] established in 1970s, with SHG intensity half that of $AgGaS_2$, and polycationic $[Sb_7S_8Br_2](AlCl_4)_3$ showing a weak SHG intensity comparable to the KDP [24]. Interestingly, recent studies show that the Sb-centered asymmetric units are packed into centrosymmetric or NCS structures via different linkage. For instance, the transition metal polyhedron, such as FeS_4 or MnS_6, links SbQ_x into centrosymmetric compounds [25, 26], while main group GaS_4 combines SbQ_x to result in a NCS compound with strong SHG effect [27]; InS_4 links SbQ_x to generate a NCS compound with the strongest Kleinman forbidden NLO coefficient to date [28, 29].

Triel or tetrel chalcogenides are also interesting systems for second-order NLO properties; examples are commercial IR NLO crystals: $AgGaQ_2$ (Q = S, Se) and $ZnGeP_2$ (ZGP) [7–11]. Newly found ternary $LiGaQ_2$ (Q = S, Se, Te) [30–32], $LiInQ_2$ (Q = S, Se), and $BaGa_4S_7$ show larger laser-induced damage thresholds [33–35]. New quaternary $Eu_2Ga_2GeS_7$ [36] exhibits powder SHG intensity 1.50 times that of $AgGaS_2$, while ternary $LaGaS_3$ [37] exhibits very weak SHG signal.

In this review, the structures and NLO properties of compounds classified according to five selected systems, Ln/T/Sb/Q, Ln/Tr/Sb/S, Ln/Tr/Tt/S, La/Ga/S, and Ba/Tr/S (Ln = lanthanide; T = transition metal; Tr = triel; Tt = tetrel; Q = chalcogenide), will be discussed.

2 Exploration on Compounds Containing Sb^{3+} with Stereochemically Active Lone Pair Electrons

2.1 Ln/T/Sb/Q Systems (Ln = Lanthanide; T = Transition Element; Q = S, Se)

A few quaternary alkali metal/lanthanide/antimony/chalcogenide compounds are known, such as $K_2Ln_{2-x}Sb_{4+x}Se_{12}$ [38], $K_2Ln_2Sb_2S_9$ [39], and $Na_9Gd_5Sb_8S_{26}$ [40], which exhibit less dense Sb/Q substructures and wider energy gap because the strong ionic bonding interactions between the cationic alkali metal and the anionic Sb/Q framework. The idea of replacement of the alkali metal by the less electropositive transition metal (T) leads to new compounds with different structure types. In Ln/T/Sb/Q systems, such as $La_4FeSb_2Q_{10}$ (Q = S, Se) and $Ln_2Mn_3Sb_4S_{12}$ (Ln = Pr–Gd) [25, 26], the properties of the component binary compounds, FeS [41] versus MnS [42], have impact on the crystal structures as well as the electronic structures.

Centrosymmetric $La_4FeSb_2S_{10}$ features unique teeter-totter $(SbS_4)_n$ chains, interconnected by LaS_8 bicapped trigonal prism into a layered motif that parallels to the *ab* plane, namely, La/Sb/S slab. Between these La/Sb/S slabs, there are La/Fe/S slabs, which are formed by the edge-sharing LaS_{10} bicapped square antiprisms and FeS_4 tetrahedra. The adjacent La/Sb/S and La/Fe/S slabs are linked via the S–S edges and apexes of LaS_8 and LaS_{10} polyhedra (Fig. 1).

Another centrosymmetric $Ln_2Mn_3Sb_4S_{12}$ can be viewed as a wavy MnS_6 octahedral layer decorated on both sides by SbS_5 square pyramid chains. The Ln^{3+} cations are located between such layers (Fig. 2). Each Mn/Sb/S layer constitutes $[Mn1S_6]_2$ dimer

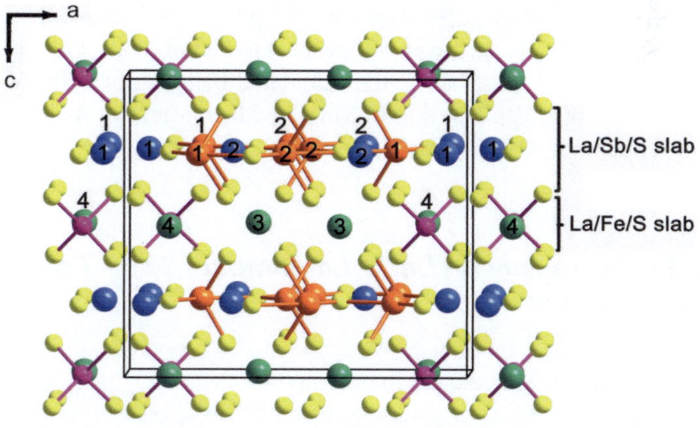

Fig. 1 View approximately along *b*-axis of $La_4FeSb_2S_{10}$. *Light blue*, La1 and La2; *Light green*, La3 and La4; *orange*, Sb; *pink*, Fe; *yellow*, S. La–S bonds are omitted for clarity. Sb–S bonds < 3.00 Å; Fe–S bonds < 2.40 Å. Reprinted with permission from [25]. Copyright 2009 American Chemical Society

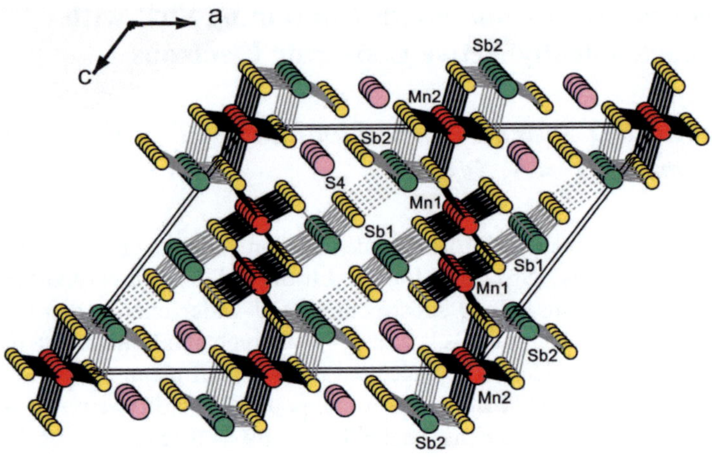

Fig. 2 The approximate (010) structure view of $Sm_2Mn_3Sb_4S_{12}$. *Pink*, Sm; *red*, Mn; *green*, Sb; *yellow*, S. Reprinted with permission from [26]. Copyright 2010 American Chemical Society

chain along [010] direction that is interconnected with $Mn2S_6$ octahedron single chain to form a layer approximately perpendicular to the *a* direction.

$La_4FeSb_2S_{10}$ has an optical bandgap of 1.00 eV, whereas $Sm_2Mn_3Sb_4S_{12}$ shows a wider gap of 1.50 eV. These values consist with their colors. The DFT calculations reveal that such a bandgap difference originates from the different contributions of Fe^{2+} and Mn^{2+} ions to the electronic structures. For both compounds, the valence bands (VB) are dominated by S 3p block. Differently, in $Sm_2Mn_3Sb_4S_{12}$, the contribution from Mn atoms near the Fermi level is almost neglectable, whereas in $La_4FeSb_2S_{10}$, the conduction band (CB) is primarily Fe 3d in character. So the optical gaps are originated from the different electronic transfer excitations, S 3p to Fe 3d in $La_4FeSb_2S_{10}$ but S 3p to Sb 5p in $Sm_2Mn_3Sb_4S_{12}$. The bandgap of $La_4FeSb_2S_{10}$ is narrower with respect to that of Sb_2S_3 (1.7–1.9 eV) [43], La_2S_3 (1.7–2.0 eV) [44], $La_7Sb_9S_{24}$ (the calculated bandgap of 1.55 eV [45]), and $K_2La_2Sb_2S_9$ (2.20 eV) [39]. But that of $Sm_2Mn_3Sb_4S_{12}$ has no obvious difference than that of ternary $La_7Sb_9S_{24}$ (Fig. 3).

2.2 *Ln/Tr/Sb/S Systems (Ln = Lanthanide; Tr = Triel, Group 13 Element)*

The above discussions indicate an important fact that SbQ_x polyhedron tends to adopt asymmetric coordination geometry, such as SbS_4 teeter-totter polyhedra or SbS_5 square pyramid. These examples also imply that the second building unit is important to the overall packing of the structure. For example, FeS_4 tetrahedron or MnS_6 octahedron both link the asymmetric SbS_x polyhedron into centrosymmetric structures [25, 26]. Recently, explorations on introduction main group metal into

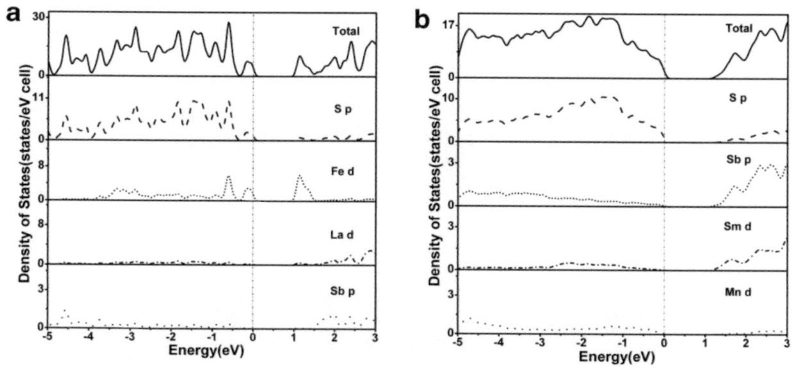

Fig. 3 Total and partial densities of state of (**a**) $La_4FeSb_2S_{10}$ and (**b**) $Sm_2Mn_3Sb_4S_{12}$

Ln/Sb/S systems give rise to interesting results. The asymmetric SbS_x building units can be sufficiently joined into NCS structures by the main group metal polyhedron. More interestingly, slight size change (GaS_4 vs. InS_4) can lead to a significantly structure rearrangement that yields excellent NLO compounds with totally different SHG origins [27, 28].

2.2.1 Ln_4GaSbS_9 (Ln = Pr, Nd, Sm, Gd–Ho): In-Phase Alignments of Asymmetric Building Units and Its Strong Nonlinear Optical Responses in Middle-IR Region

Ln_4GaSbS_9 (Ln = Pr, Nd, Sm, Gd–Ho) [27] crystallizes in orthorhombic NCS space group, *Aba*2 (No.41). The structure features infinite single anionic chains of $\{[(Ga_2S_6)(Sb_2S_5)]^{10-}\}_\infty$ that are constructed from bimetallic asymmetric Sb_2S_6 units and dimeric Ga_2S_7 tetrahedra. These parallel infinite chains are separated by Ln^{3+} cations and discrete S^{2-} anions.

Figure 4 shows the packing of these chains in a NCS pseudo-layer motif perpendicular to the *c*-axis. Another view down *c*-axis showing the overlay of two such pseudo-layers is in Fig. 5a. The *c* glide plane at (1/4, *y*, *z*) relates the lower chains ($0 < c < 0.5$) to the upper chains ($0.5 < c < 1.0$). This NCS packing allows the in-phase alignment of the dipoles of both Sb_2S_6 and Ga_2S_7 units as indicated in Fig. 5b, which is mainly responsible for the strong SHG response.

Sm_4GaSbS_9 shows strong powder SHG effects at 2.05 μm with the intensity about 3.8 times that of commercial $AgGaS_2$ (Fig. 6). The transparency ranges in 1.75–25 μm is comparable to 0.60–25 μm of powder $AgGaS_2$. The bandgap of 2.23 eV is close to that of $AgGaS_2$ (2.62 eV) [9, 46].

The VASP calculations visualize stereochemically active lone pair electrons on Sb and reveal that the driving force of the formation of such lone pairs is the Sb–S bonding interaction. The fundamental band edge excitation originates from transitions involving the filled primarily S 3p states in VB and the empty Sm 5d states in CB.

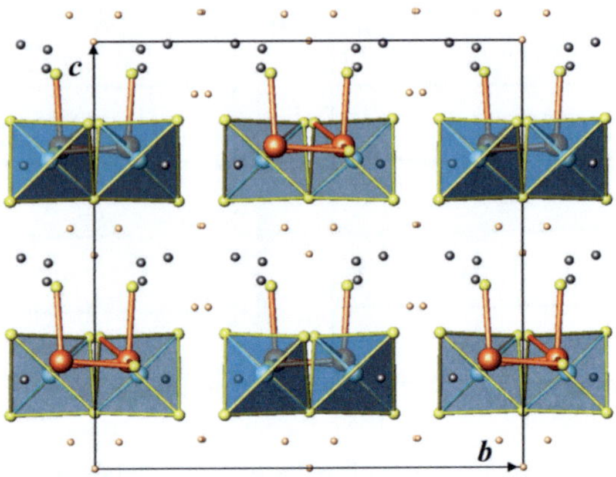

Fig 4 Structure of orthorhombic Sm_4GaSbS_9 viewed down the *a*-axis with unit cell marked. The Sm–S bonds are omitted for the sake of clarity. *Blue*: Ga; *orange*: Sb; *yellow*, S; *pink*: interlayer S7, S8, S9, and S10; *black*: Sm; blue tetrahedron: GaS_4 tetrahedron. Anionic $\{[(Ga_2S_6)(Sb_2S_5)]^{10-}\}_\infty$ chains are arranged in a pseudo-layer motif perpendicular to the *c*-axis. Reprinted with permission from [27]. Copyright 2011 American Chemical Society

2.2.2 Ln_4InSbS_9 (Ln = La, Pr, Nd): Chiral Arrangement of Asymmetric Building Units and Its Strong Kleinman Forbidden Second-Harmonic Generation

Ln_4InSbS_9 (Ln = La, Pr, Nd) crystallizes in space group, $P4_12_12$ (No.92), or its enantiomorph, $P4_32_12$, (No.96) and is characterized by infinite 2_1 helical chains of $[In_2Sb_2S_{11}{}^{10-}]_\infty$ propagating along the *c* direction that are separated by isolate Ln^{3+} cations and discrete S^{2-} anions [28]. These chains are further packed around the 4_1 or 4_3 helical axes located at (1/2, 0, *z*), (0, 1/2, *z*), (1/2, 1, *z*), and (1, 1/2, *z*), respectively (Fig. 7a) [28]. Such chains are built by dimeric teeter-totter Sb_2S_6 polyhedra and dual apex-shared In_2S_7 tetrahedra (Fig. 7b), which are reminiscence of the chains in Ln_4GaSbS_9 [27], but differ in that Sb_2S_6 and In_2S_7 are arranged around a twofold screw axis, thus the neighboring dimmers are in opposite direction (Fig. 7c), while the neighboring Sb_2S_6 or Ga_2S_7 in the latter are in-phase arranged (Fig. 5b).

SHG measurements reveal that La_4InSbS_9 displays SHG signals about 1.50 times that of commercial $AgGaS_2$ (Fig. 8). Moreover, the SHG intensity increases with the particle size and is saturated at a maximum value indicating a type-I phase matchable behavior (Fig. 9). The optical bandgap is estimated to be around 2.07 eV, which is comparable with that of $AgGaS_2$ (2.62 eV) [9, 46], implying that this compound has suitable laser-damage threshold for NLO applications. Moreover, the transparency is wide in the range of 1.00–25.00 μm and excellent thermal stability up to 765°C. So La_4InSbS_9 may be a potential candidate for the application in the middle and far-IR region.

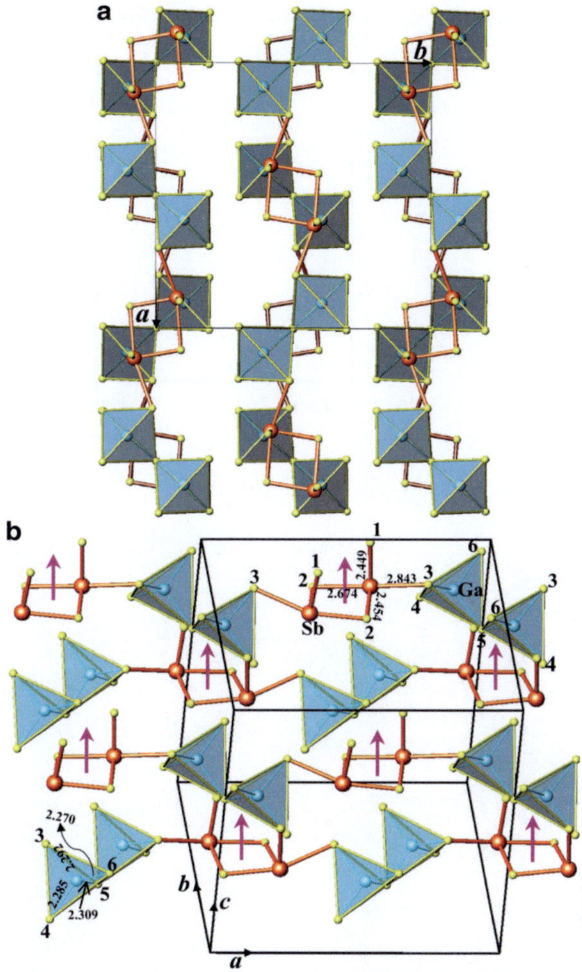

Fig 5 **(a)** The noncentrosymmetric packing of the $\{[(Ga_2S_6)(Sb_2S_5)]^{10-}\}_\infty$ chains in Sm_4GaSbS_9 viewed down the *c*-axis. *Blue*: Ga; *orange*: Sb; *yellow*, S. Chains with *dark* color are at $0.5 < c < 1.0$, and chains with *bright* color are at $0 < c < 0.5$. The interlayer Sm and S ions are omitted for the sake of clarity. Reprinted with permission from [27]. Copyright 2011 American Chemical Society. **(b)** A pseudo-layer of separated infinite $\{[(Ga_2S_6)(Sb_2S_5)]^{10-}\}_\infty$ chains. The *pink* arrows show the in-phase alignment of such chains. Reprinted with permission from [27]. Copyright 2011 American Chemical Society

La_4InSbS_9 crystallizes in space group, $P4_12_12$, 422 point group, whose space group symmetry requires two nonvanishing tensors of second-order susceptibilities to follow the equation $d_{14} = -d_{25}$, which have to be zero under the restriction of Kleinman's symmetry. Thus, La_4InSbS_9 should exhibit null SHG response by Kleinman's symmetry [29]. The assumption of the Kleinman's symmetry is ideal that the medium is dispersionless; however, under the real experimental condition,

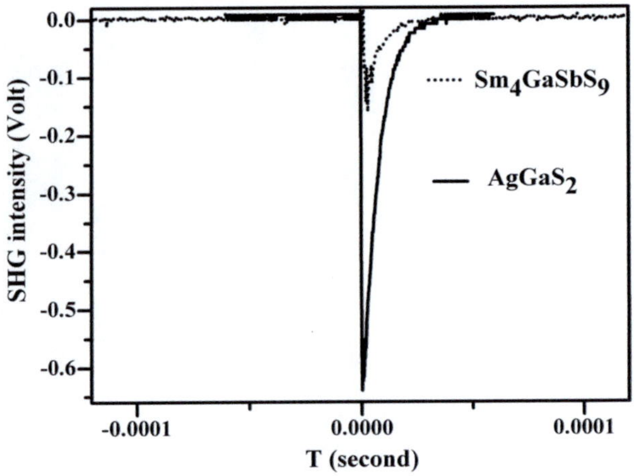

Fig 6 Oscilloscope traces of the second-harmonic-generated signals of Sm_4GaSbS_9 and $AgGaS_2$

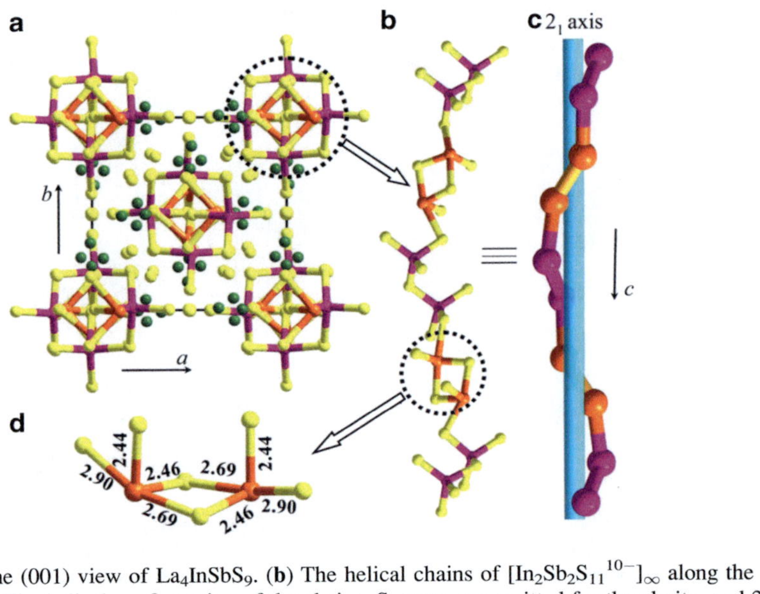

Fig 7 (**a**) The (001) view of La_4InSbS_9. (**b**) The helical chains of $[In_2Sb_2S_{11}{}^{10-}]_\infty$ along the *c* direction. (**c**) The helical configuration of the chains, S atoms are omitted for the clarity, and 2_1 screw axis is visualized. (**d**) The local coordination of the dimeric teeter-totter Sb_2S_6. *Green*, La; *pink*, In; *orange*, Sb; *yellow*, S. Reprinted with permission from [28]. Copyright 2012 American Chemical Society

dispersion always exists, which can damage the validity of Kleinman's symmetry so as to lead to very weak SHG response [47–54]. For instance, α-TeO_2 crystallizing in $P4_12_12$ exhibits a weak d_{14} (1.328 μm) ≈ 0.36 pm/V [47]. Only if the second-harmonic frequency is close to an absorption bandgap ($2h\omega \approx Eg$) will the electronic exciton resonance be expected, which can significantly enhance the SHG effect [55, 56]. For example, α-TeO_2 shows a strong Kleinman forbidden d_{14} (0.659 μm) ≈ 4.32 pm/V, which is 10 times larger than that at 1.328 μm, because

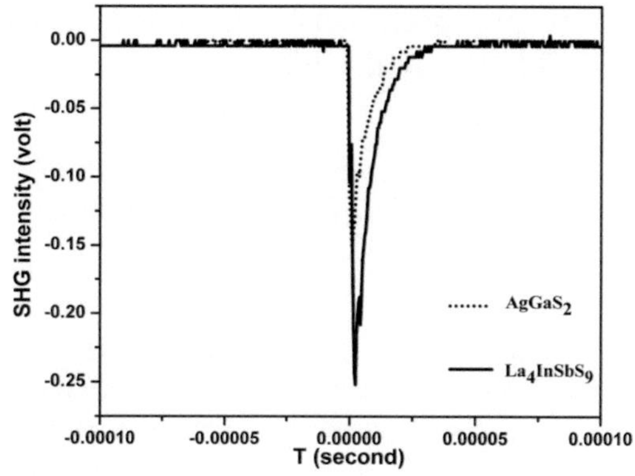

Fig 8 Oscilloscope traces of the second-harmonic-generated signals of La_4InSbS_9 and $AgGaS_2$

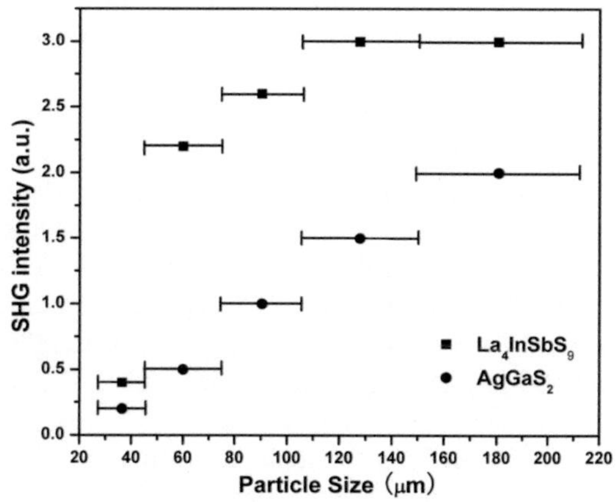

Fig 9 Particle-size dependence of second-harmonic intensity for La_4InSbS_9 and benchmark $AgGaS_2$

the second-harmonic wavelength 0.33 μm (~ 3.76 eV) is very close to its optical bandgap (3.75 eV) [47, 57, 58].

But, this is not the case for La_4InSbS_9. Under the measurement condition, the second-harmonic wavelength is about 1.00 μm (about 1.2 eV), which differs significantly from the optical bandgap (2.07 eV); therefore, electronic resonance will not be expected. The second-order NLO susceptibility coefficient calculations based on length-gauge formalism derived by Aversa and Sipe further confirm this. The calculations generate null static SHG coefficients under the zero-frequency limit [59]. Consequently, the second-order NLO coefficients of 11 possible configurations of La_4InSbS_9 simulated by the ab initio molecular dynamic calculations at 300 K have been calculated [60]. Nine out of eleven configurations have at least a NLO coefficient close to 15 pm/V. This value is roughly 1.2 times d_{36} (10.6 μm) of $AgGaS_2$ (13 pm/V) [61] and is in agreement with the experimental observations. These results strongly suggest that the lattice thermal vibrations induce the structure configurational variation, which may be responsible for the strong SHG effect of La_4InSbS_9.

3 Exploration on Compounds Containing TrS_4 and TtS_4 Tetrahedral Units

3.1 *Ln/Tr/Tt/S Quaternary Systems (Ln = Lanthanide; Tr = Triel, Group 13 Element; Tt = Tetrel, Group 14 Element)*

β-$LaGaS_3$ is a NCS compound based on the wavy GaS_4 tetrahedron chains separated by La^{3+} cations [37] and exhibits very weak SHG effect. Introducing Ge to Ln/Ga/S system results in two new thiogermanate: $La_2Ga_2GeS_8$ and $Eu_2Ga_2GeS_7$ [36]. $La_2Ga_2GeS_8$ contains infinite chains of corner-sharing GaS_4 tetrahedra, interconnected via GeS_4 tetrahedra into $[Ga_2GeS_8]^{6-}$ layers that are separated by La^{3+} (Fig. 10a). $Eu_2Ga_2GeS_7$ comprises corner-sharing diametric Ga_2S_7, which is further interconnected by GeS_4 tetrahedra into $[Ga_2GeS_7]^{4-}$ layers that are separated by Eu^{2+} ions (Fig. 10b). While $Li_2Ga_2GeS_6$ [62] is a 3D framework built by GaS_4 and GeS_4 tetrahedra embedding tunnels along the *c*-axis, in which Li^+ ions locate (Fig. 10c). These three compounds represent a nice group of compound differing only in the oxidation state of cation, La^{3+}, Eu^{2+}, and Li^+. The oxidation state of the cation affects the total numbers of S atoms so as to influence the overall crystal-packing pattern. The number of terminal S atom are 4, 2, and 0 in $La_2Ga_2GeS_8$, $Eu_2Ga_2GeS_7$, and $Li_2Ga_2GeS_6$, respectively. According to $2n\mu_2{-}S + n\mu_t{-}S = 12$, the number of terminal and bridging S atoms per formula can be calculated. It can deduce that $n\mu_2{-}S = 6$, $n\mu_t{-}S = 0$ for $Li_2Ga_2GeS_6$; $n\mu_2{-}S = 5$, $n\mu_t{-}S = 2$ for $Eu_2Ga_2GeS_7$; and $n\mu_2{-}S = 4$; $n\mu_t{-}S = 4$ for $La_2Ga_2GeS_8$. These calculated results are identical with the data observed directly from the single crystal structure refinement results. As the number of terminal S increases, the packing of the anionic moiety is looser; therefore, the number of terminal S can be regarded as a structure indicator.

Interestingly, $Eu_2Ga_2GeS_7$ reveals an SHG response of about 1.5 times that of $AgGaS_2$, and $La_2Ga_2GeS_8$ shows a very weak response. Regarding to the origin of the SHG intensity, there are different opinions. One theoretical study considers the SHG coefficient of $LiGaQ_2$ (Q = S, Se, Te) coming from the contribution of GaQ_4 tetrahedron [30–32]. Another extensive study on $La_2Ga_2GeS_8$, $Eu_2Ga_2GeS_7$, and $Li_2Ga_2GeS_6$ finds that the SHG coefficients originate from the electronic transitions from the S 3p states to the La/Eu/Li − S, Ga − S, and Ge − S antibonding states [36]. Thus, except the anionic moiety, cation also contributes to the SHG activity.

3.2 *Ba/Tr/S Ternary Systems (Tr = Triel, Group 13 Element)*

This system contains a large number of compounds including $Ba_2In_2S_5$ [63], $BaIn_2S_4$ [64], $Ba_2Ga_2S_5$ [65], $Ba_3Ga_2S_6$ [66], $Ba_4Ga_2S_7$ [66], $Ba_5Ga_2S_8$ [67], and

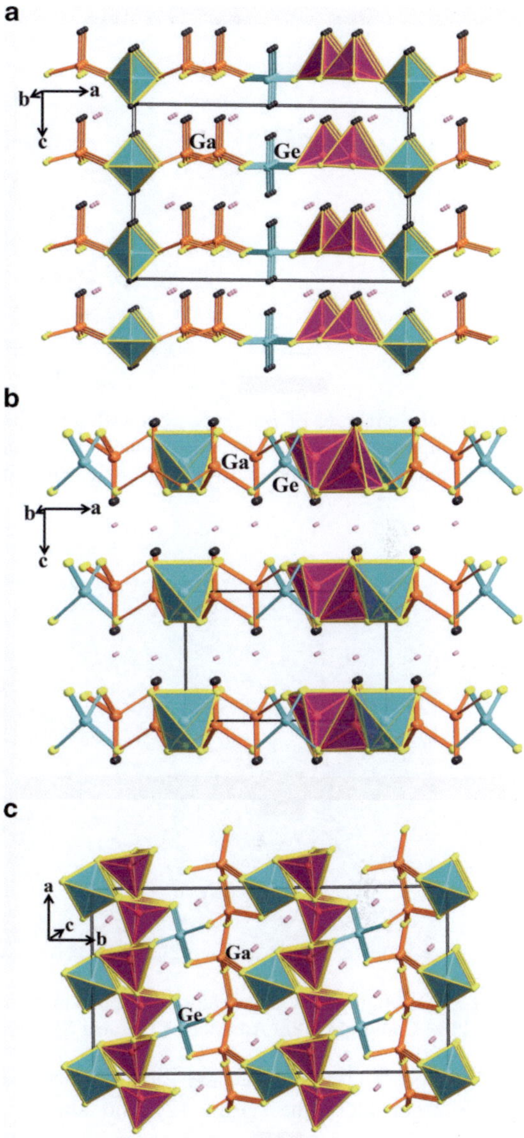

Fig 10 The structures of (**a**) $La_2Ga_2GeS_8$, (**b**) $Eu_2Ga_2GeS_7$, and (**c**) $Li_2Ga_2GeS_6$ with unit cells outlined. *Yellow*: μ_2–S; *black*: μ_t–S; *orange*: Ga; *light green*: Ge; *pink*: $La^{3+}/Eu^{2+}/Li^{+}$. *Light green* tetrahedron: GeS_4; *pink* tetrahedron: GaS_4. Reprinted with permission from [36]. Copyright 2011 American Chemical Society

$BaGa_2S_4$ [68], and their structures are based on the same primary building unit, TrS_4 tetrahedron. Several compounds, $Ba_{12}In_4S_{19}$, $Ba_4In_2S_8$, and $Ba_4Ga_2S_8$, involve disulfide S_2^{2-} anion [69]. $Ba_{12}In_4S_{19} \equiv (Ba^{2+})_{12}(In^{3+})_4(S^{2-})_{17}(S_2)^{2-}$ shows a unique long period-stacking structure of a combination of monometallic InS_4 tetrahedra, linear dimeric In_2S_7 tetrahedra, disulfide S_2^{2-} anions, and isolated sulfide S^{2-} anions that are further enveloped by Ba^{2+} cations (Fig. 11). Interestingly, the CASTEP calculations show that the disulfide S_2^{2-} anions mainly determine the bandgaps.

Fig 11 The structure of $Ba_{12}In_4S_{19}$, viewed approximately along *c*-axis. *Black*: Ba^{2+}; *yellow*, S^{2-} anion; *light blue* tetrahedra: monometallic $In5S_4$; *purple* tetrahedra: linear dimeric In_2S_7 and monometallic $In1S_4$. Reprinted with permission from [69]. Copyright 2011 American Chemical Society

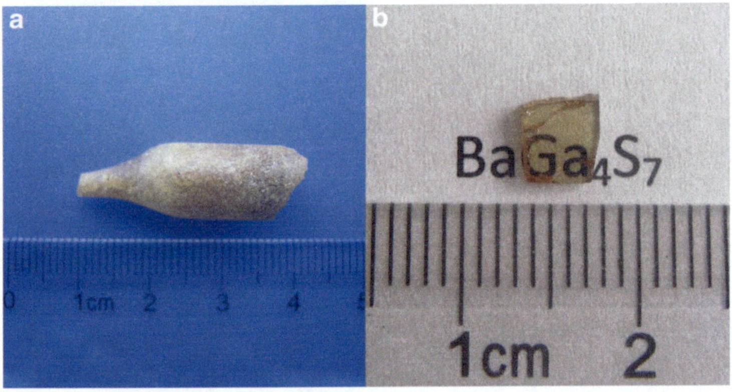

Fig 12 (**a**) As-grown BGS crystal; (**b**) polished piece of BGS crystal. Reprinted with permission from [33]. Copyright 2009 American Chemical Society

New NLO crystal $BaGa_4S_7$ [33] has been successfully grown by a Bridgman–Stockbarger technique (Fig. 12) and shows a wide optical transparent region (0.35–13.70 μm), high second-order susceptibility coefficients d_{33} (12.6 pm/V), and high laser-damage threshold (1.2 J/cm^2 at 1.064 μm and a 15 ns pulse width). This sheds useful light on the large size single crystal growth for IR NLO materials.

4 Future Prospects

In conclusion, 23 new multinary chalcogenides belonging to the five selected systems, Ln/T/Sb/Q, Ln/Tr/Sb/S, Ln/Tr/Tt/S, La/Ga/S, and Ba/Tr/S (Ln = lanthanide; T = transition metal; Tr = triel; Tt = tetrel; Q = chalcogenide), are summarized regarding their crystal and electronic structures, SHG properties, and their possible origins. In these systems, to obtain a NCS compounds, the involvement of at least an asymmetric building unit, such as SbS_x polyhedron, is necessary. The choice of the second building unit is also important, for example, GaS_4 or InS_4 is found to be more effective than FeS_4 or MnS_6 in producing NCS compounds. Besides, the size of main group metals, InS_4 versus GaS_4, affects the overall crystal structures that exhibit SHG activities coming from totally different origins. Although the combination of two types of asymmetric units, such as SbS_x and TrS_4 or TtS_4 not only gives rise to a rich structural chemistry but also yields many NCS compounds with excellent IR NLO performance, the formation of NCS compound is not guaranteed. Therefore, the rational design of NCS compounds with good SHG properties remains a great challenge, and exploration in this field is always the first step!

Acknowledgment This research was supported by the National Natural Science Foundation of China under projects (20973175, 21171168), 973 Program (2010CB933501), and the NSF of Fujian Province (2011J05039).

References

1. Halasyamani PS, Poeppelmeier KR (1998) Chem Mater 10:2753
2. Chen C, Wu B, Jiang A, You G (1985) Sci Sin B28:235
3. Chen C, Wu Y (1989) J Opt Soc Am B 6:616
4. Rhodes MA, Woods B, DeYoreo JJ, Roberts D, Atherton LJ (1995) Appl Opt 34:5312
5. Kato K (1991) IEEE J Quant Electron 27:1137
6. Hagerman ME, Poeppelmeier KR (1995) Chem Mater 7:602
7. Chemla DS, Kupecek PJ, Robertson DS, Smith RC (1971) Opt Commun 3:29
8. Boyd GD, Kasper H, McFee JM (1971) IEEE J Quant Electron QE-7:563
9. Bhar GC, Smith RC (1972) Phys Status Solidi A 13:157
10. Boyd GD, Kasper HM, McFee JH, Storz FG (1972) IEEE J Quant Electron 8:900
11. Boyd GD, Buehler E, Storz FG (1971) Appl Phys Lett 18:301
12. Jackson AG, Ohmer MC, LeClair SR (1997) Infrared Phys Technol 38:233
13. Ra HS, Ok KM, Halasyamani PS (2003) J Am Chem Soc 125:7764
14. Goodey J, Broussard J, Halasyamani PS (2002) Chem Mater 14:3174
15. Chi EO, Ok KM, Porter Y, Halasyamani PS (2006) Chem Mater 18:2070
16. Sivakumar T, Chang HY, Baek J, Halasyamani PS (2007) Chem Mater 19:4710
17. Kong F, Huang SP, Sun ZM, Mao JG, Cheng WD (2006) J Am Chem Soc 128:7750
18. Jiang HL, Huang SP, Fan Y, Mao JG, Cheng WD (2008) Chem Eur J 14:1972
19. Zhang M, Sheng TL, Wang X, Wu XT (2010) Cryst Eng Commun 12:73
20. Zhang M, Sheng TL, Huang XH, Fu RB, Wang X, Hu SM, Xiang SC, Wu XT (2007) Eur J Inorg Chem 2007:1606
21. Wang X, Sheng TL, Hu SM, Fu RB, Wu XT (2009) Inorg Chem Commun 12:399
22. Wang X, Sheng TL, Chen JS, Hu SM, Fu RB, Wu XT (2009) J Mol Struct 936:142
23. Gandrud WB, Boyd GD, Mcfee JHN (1970) Appl Phys Lett 16:59
24. Zhang Q, Chung I, Jang JI, Ketterson JB, Kanatzidis MG (2009) J Am Chem Soc 131:9896
25. Zhao HJ, Li LH, Wu LM, Chen L (2009) Inorg Chem 48:11518
26. Zhao HJ, Li LH, Wu LM, Chen L (2010) Inorg Chem 49:5811
27. Chen MC, Li LH, Chen YB, Chen L (2011) J Am Chem Soc 133:4617
28. Zhao HJ, Zhang YF, Chen L (2012) J Am Chem Soc 134:1993

29. Kleinman DA (1962) Phys Rev 126:1977
30. Isaenko L, Yelisseyev A, Lobanov S, Titov A, Petrov V, Zondy JJ, Krinitsin P, Merkulov A, Vedenyapin V, Smirnova J (2003) Cryst Res Technol 38:379
31. Isaenko L, Krinitsin P, Vedenyapin V, Yelisseyev A, Merkulov A, Zondy JJ, Petrov V (2005) Cryst Growth Des 5:1325
32. Bai L, Lin ZS, Wang ZZ, Chen CT (2008) J Appl Phys 103:083111–1
33. Lin XS, Zhang G, Ye N (2009) Cryst Growth Des 9:1186
34. Boyd GD, Kasper HM, McFee JH (1973) J Appl Phys 44:2809
35. Isaenko L, Yelisseyev A, Lobanov S (2002) J Appl Phys 91:9475
36. Chen MC, Li P, Zhou LJ, Li LH, Chen L (2011) Inorg Chem 50:12402
37. Li P, Li LH, Chen L, Wu LM (2010) J Solid State Chem 183:444
38. Chen JH, Dorhout PK (1997) J Alloys Compd 249:199
39. Choi KS, Hanko JA, Kanatzidis MG (1999) J Solid State Chem 147:309
40. Park S, Kim SJ (2001) J Solid State Chem 161:129
41. Gosselin JR, Townsend MG, Trembly RJ (1976) Solid State Commun 19:799
42. Kan SH, Felner I, Banin U (2001) Israel J Chem 41:55
43. Shutov SD, Sobolev VV, Popov YV, Shestatskii SN (1969) Phys Status Solidi 31:K23
44. Gschneidner KA, Beaudry BJ, Takeshita T, Eucker SS, Taher SMA, Ho JC (1981) Phys Rev B 24:7187
45. Assoud A, Kleinke KM, Kleinke H (2006) Chem Mater 18:1041
46. Setzler SD, Schunemann PG, Pollak TM, Ohmer MC, Goldstein JT, Hopkins FK, Stevens KT, Halliburton LE, Giles NC (1999) J Appl Phys 86:6677
47. Levine BF (1973) IEEE J Quant Electron QE-9:946
48. Singh S, Bonner WA, Van Uitert LG (1972) Phys Lett 38A:407
49. Chemla DS, Jerphagnon J (1972) Appl Phys Lett 20:222
50. Porter Y, Halasyamani PS (2001) Chem Mater 13:1910
51. Guo SP, Guo GC, Wang MS, Zou JP, Zeng HY, Cai LZ, Huang JS (2009) Chem Commun 4366
52. Crane GR, Bergman JG (1976) J Chem Phys 64:27
53. Okada M, Ieiri S (1971) Phys Lett 34A:63
54. Franken PA, Ward JF (1963) Rev Mod Phys 35:23
55. Wynne JJ (1971) Phys Rev Lett 27:17
56. Zhang XQ, Tang ZK, Kawasaki M, Ohtomo A, Koinuma H (2003) J Phys Condens Matter 15:5191
57. Levine BF, Miller RC (1975) Phys Rev B 12:4512
58. Al-Kuhaili MF, Durrani SMA, Khawaja EE, Shirokoff J (2002) J Phys D: Appl Phys 35:910
59. Rashkeev SN, Lambrecht WRL (2001) Phys Rev B 63:165212
60. Nosé S (1984) J Chem Phys 81:511
61. Dmitriev VG, Gurzadyan GG, Nikogosyan DN (1999) Handbook of nonlinear optical crystals, 3rd edn. Springer, New York
62. Kim Y, Seo IS, Martin SW, Baek J, Halasyamani PS, Arumugam N, Steinfink H (2008) Chem Mater 20:6048
63. Eisenmann B, Hofmann A (1990) Z Anor Allg Chem 580:151
64. Eisenmann B, Jakowski M, Klee W, Schafer H (1983) Rev Chim Miner 20:255
65. Eisenmann B, Jakowski M, Schafer H (1983) Z Naturforsch 38:1581
66. Eisenmann B, Jakowski M, Schafer H (1984) Rev Chim Miner 21:12
67. Eisenmann B, Jakowski M, Schafer H (1984) Z Naturforsch 39:27
68. Ivanov-Emin BN, Ivlieva VI, Filatenko LA, Sarabiya MG, Kaziev GZ, Zaitsev BE (1980) Russ J Inorg Chem 25:515
69. Liu JW, Wang P, Chen L (2011) Inorg Chem 50:5706

Index

MIX
Papier aus verantwortungsvollen Quellen
Paper from responsible sources
FSC® C105338
FSC
www.fsc.org

If you have any concerns about our products,
you can contact us on
ProductSafety@springernature.com

In case Publisher is established outside the EU,
the EU authorized representative is:
Springer Nature Customer Service Center GmbH
Europaplatz 3, 69115 Heidelberg, Germany

Printed by Libri Plureos GmbH
in Hamburg, Germany